W9-BYP-683

WHAT ARE THEY SAYING ABOUT
GENETIC ENGINEERING?

What Are They Saying About Genetic Engineering?

Thomas A. Shannon

PAULIST PRESS
New York / Mahwah

The Publisher gratefully acknowledges the use of "'Making Babies' Revisited" by Leon Kass. Reprinted with permission of the author from: THE PUBLIC INTEREST, No. 54 (Winter, 1979), pp. 32-60. © 1979 by National Affairs, Inc.

Library of Congress
Catalog Card Number: 85-61751

ISBN: 0-8091-2743-1

Published by Paulist Press
997 Macarthur Boulevard
Mahwah, N.J. 07430

Printed and bound in the
United States of America

Contents

To
my mother Clara J. Shannon
and
my mother-in-law Imogene F. Haenn
with gratitude for all they have given me

1
Introduction

A. Purpose of the Book

The overall purpose of the book is to describe and examine several contemporary developments in genetics which raise a variety of ethical and social problems. To do this I will describe several of these developments, indicate some of the thematic ethical issues to which they point, and describe some of the responses that have been made to them. The book will present the contours of many of the debates and will specifically examine several different problems within the area of genetic engineering itself.

B. Some Definitions

Genetics is the division of biology that focuses primarily on the genes or the units of the chromosome that determine one's inheritance. Occasionally in bioethical discussions, the term genetics is used as a shorthand reference to many of the disciplines and developments within the life sciences. Although I will occasionally use the word in that broad sense, I will primarily use it in the technical sense of the study of genes and the application of that knowledge in a variety of experimental and clinical situations.

The genes are the basic blueprint or plan for heredity, the program which helps specify how the organism develops. The genes are made up of segments of deoxyribonucleic acid, commonly abbreviated as DNA, in which four chemical subunits are united in an incredible variety of ways. These four chemicals, commonly abbreviated as A, G, C, and T, form the alphabet or code which instructs the cell in making proteins and ultimately in building up the entire organism. Within recent years geneticists have learned more and more about the composition of this alphabet, the processes and ways by which the message is written and communicated, and methods of decoding it. Such a growing understanding of the genetic code has occasioned many discussions of the problems of the applications and implications of genetics that we will present in this book.

The term genetic engineering has a similar narrow and broad meaning. Strictly speaking, it refers to specific technical interventions in the structure of the gene for a variety of purposes, including, but not limited to, removing a deleterious or harmful gene, changing the genetic structure of a particular organism, or enhancing a particular genetic capacity. Genetic engineering is a specific intervention into the actual gene structure itself. In a broader sense, and the way it is most often used, genetic engineering refers to the possibility of not only designing our descendants, but also manipulating the entire ecosystem for a variety of purposes. More specifically, though, genetic engineering in a broad sense refers to technologies such as in vitro (or test tube) fertilization, cloning (the artificial reproduction of an identical twin), recombinant DNA research, and a variety of other applications of this knowledge of the gene structure, together with the social, political, and ethical dimensions of these applications.

One of the major spin-offs of our understanding of genetics has been the development of programs of genetic screening in which potential carriers of a variety of genetic diseases can be identified and informed of their situation so that they can make a more informed reproductive decision. Such screening programs involve an identification of a target population and an examination of the chromosomes of the individuals in that population to determine whether or not they contain deleterious or harmful genes. On the basis of information thus obtained, further genetic counseling can be obtained if desired.

Another area of growing importance is genetic engineering by removing or replacing an individual's harmful genes so that a disease will not occur or can be corrected. Such developments offer the possibility of intervening directly into an individual's genetic structure for reasons of therapy or the achievement of personal or social desires. The technology of recombinant DNA also makes it possible to envision the development of a new species of life forms.

Connected with this is a variety of developments in birth technologies in which the external fertilization of a fetus is now possible; embryo transplants in which an embryo can be transplanted from one uterus to another; surrogate parenting in which one person conceives and carries a child on behalf of another person. This externalization of fertilization permits a greater potential for manipulation of the fetus than had been possible.

Another area related to developments in genetics is not really a technology but an initial formulation of an academic discipline: sociobiology. This is the systematic study of the biological bases of all forms of social behavior in all organisms, including human beings. This new discipline is important because of the far reaching questions it raises about the origins of various kinds of human behavior, especially altruism and freedom. The implications of the questioning of these valued forms of human behavior raise significant questions that need to be addressed in at least a preliminary fashion.

C. General Ethical Questions

Before I examine specific topics, let me present several thematic questions that cut across the individual topics to be examined in the remaining chapters.

1. On what basis does one intervene into the genetic structure of an individual?

Two basic motivations present themselves. The first is therapeutic. That is, one wishes to make an intervention into the biology of an individual to correct a gene that is defective or to repair it so that the individual will be restored to a normal or appropriate level of health. A second motivation is eugenic, in which one intervenes for the purpose of enhancing or developing specific characteristics that are perceived to be desirable. There is a long history of such eugenic interventions in animals and many argue that within certain limits there is no reason why such genetic selectivity could not also be useful in humans.

Each of these motivations raises a variety of social and ethical problems. The therapeutic interventions present some problems with respect to risk and the consequences of such an intervention. But these can be understood, at least initially, within the moral framework of our traditional understanding of medicine. Eugenic interventions, focusing as they do on enhancing some behaviors or controlling others, raise a variety of issues that we have only recently begun seriously thinking about. Development of intelligence, gender selection and issues of domination and social status are examples of some of these issues.

Chapters 2, 4, and 6 will examine this question in more detail.

2. On the basis of what criteria would one intervene?

Many criteria present themselves as possible candidates for intervention: behavioral, social, medical, political, and economic. Each of these criteria carries with it significant value choices and ethical implications which need to be thoroughly examined as different policies are put into place.

For example, there were a variety of medical and social criteria used to set up genetic screening programs in the late '60s and early '70s. The motivations behind them were designed to benefit the target population. On the other hand, some of the recipients of these programs saw them as discriminatory and, in fact, potentially genocidal because of their impact on the target population. Thus while the motivation for establishing a set of criteria may be based primarily on the value of beneficence, one still needs to look at the implications of and perceptions about such a program.

Chapters 4, 6 and 9 examine this question further.

3. Who sets the criteria?

Assuming that intervention is possible and that criteria can be developed, the critical question yet remains: who is responsible for establishing such criteria? Scientists, physicians, politicians, citizens, the recipients of the technology, or the relatives of the recipients are all possible candidates. One can establish a whole variety of reasons for selecting one member of these possible candidates as the most appropriate decision maker or one to establish the criteria.

One of the major problems that needs to be avoided is falling into the fallacy of the generalization of expertise. That is the problem of an individual's having substantive qualifications in one area, for example genetics, and then extending those credentials well beyond their capacity. This individual could then also become responsible for setting social policy. While it does not follow that those with expertise in one area should not be decision makers, we must also ensure that we do not assume these people to be decision makers simply because they have expertise in some area.

Chapters 2, 3, and 10 examine this particular question from various aspects.

4. What are the risks of intervention?

Potential risks of intervention can be thought of on an individual, social and environmental scale. With respect to the individual, many things can go wrong when an intervention is made and those potentials should be thoroughly examined beforehand. From a social point of view, we need to attend to the rationale for selecting a variety of values that may be unique or appropriate only to one small segment of a particular society. The issue here is that while some values may be good for some people, it is not necessarily the case that these particular values will be appropriate for all individuals.

Another issue that needs to be carefully examined is the environmental implications of genetic engineering. Unfortunately this is an area in which we have the least amount of knowledge and, because of this, we have the most difficulty in predicting possible risks to the environment. On the one hand, it is possible to argue for a totally conservative vision which says that nature knows best and that we should not attempt to put new forms of life into the environment. On the other hand, we know that much of what occurs in nature causes severe problems for individuals. Consequently, we do have a well justified tradition of intervention into nature. We need to be concerned about thinking through, insofar as possible, what kinds of implications would come from specific interventions into the environment and what risks that might bring.

Chapters 3, 4 and 6 examine the details of that particular question.

5. What are the benefits and who receives them?

As indicated, the question of benefits always has a two edged dimension. On the one hand, we need to be reasonably certain that what we are doing is actually a benefit. For example, an intervention may in fact ensure the survival of an individual, but it may also leave that individual at a rather low quality of life. What was initially perceived as a benefit ultimately turned into a serious burden. We need to ensure that what we intend as a benefit actually does emerge as one.

We also need to consider who is going to receive these benefits. Is it only the people who can pay for them; will it be only those willing to undergo a variety of experimental procedures; will it be anyone who presents a need for the service? There are serious questions of cost involved in this with respect to who pays for the benefit and also whether such specialized forms of treatment can be available to all on a routine basis.

6. For what future are we planning?

Clearly, in many of the scenarios of genetic engineering, various forms of gene therapy, and birth technologies that are being proposed and developed, there is a vision of a particular future. This future, obviously, is based on certain values, preferences, and choices. At some point we need to ask ourselves whether this is the future that we want and is this a future that will respect and enhance the rights of those who live in that future.

Obviously, this is a very difficult question but it is a question that needs to be raised continuously because the American pragmatic tradition has generally not asked these kinds of questions. Typically we go ahead with what we can do and only after we are involved in a situation do we ask ourselves whether or not we should be there. By that time it is generally too late. Resources have been committed. Research has begun. Expectations have been raised. Problems or potential problems are secondary to these overriding issues.

Chapters 4, 7, 8, and 10 examine further implications of this question.

7. What is the relationship between genetics and behavior?

The traditional question of nature versus nurture is a very sticky one because it is almost impossible to design an ethical experiment that would isolate the relevant factors so that the precise relationship could be evaluated. On the other hand, in talking about genetic engineering we need to be reminded almost constantly that genes operate within a biological and social environment. To speak of genetic engineering without also remembering this context is to cast the question in an inappropriate fashion. If we think only of the genetic dimensions and do not also attend to the critical social and environmental issues, we will have problems in developing appropriate therapies and also in attempting to make the appropriate intervention.

Chapters 2, 4, 5, and 7 pursue this problem further.

These are several of the issues that will be attended to through the remaining chapters of this book. The issues are very complex and my purpose is to provide thematic discussions of them and to indicate several of the positions that are currently being argued and to provide some orientation toward the solutions that are being offered.

2
Science and Society

As a result of developments and applications in science such as nuclear energy, the breaking of the genetic code, and hypotheses about the relation between biology and behavior, we know more clearly than ever before that social interests and needs are as much a part of science as is the desire to understand nature. Our increased capabilities give us new powers which in turn create the opportunity to redesign our world and ourselves.

In this chapter, we will examine three areas of interrelation of science and society: perceptions of the relation, different models of scientific research, and the social responsibilities of the scientist.

A. Perceptions of Science

Two viewpoints on how science is perceived are provided by contemporary commentators. Pope John Paul II, in his first encyclical letter, *Redemptor Hominis*, suggested that humans are becoming afraid of what they produce because they perceive these objects could turn against them. He emphasized the growing fear that our products can become the means and instruments for self-destruction. Why is it, the Pope asks, that this power has turned against humans and produced a state of disquiet, of fear and menace? A partial answer is that people now perceive themselves not as masters or guardians of the world but as its exploiters and destroyers. Coupled with this is an uncertainty of whether or not our products make our life more human and, therefore, more worthy of persons.[1]

Daniel Callahan gives a contrasting perspective.[2] He reports that current developments in genetic engineering, broadly speaking, suggest that both the scientific community and the general public are more prepared than ever to go ahead with new developments and applications.

Callahan indicates that there has been a typical reaction of wonder and excitement in both the scientific and public media whenever a major breakthrough has been discovered. He concludes that society is fascinated by scientific progress and technological applications of new insights into the processes of nature.

Callahan suggests that this attitude prevails because no really persuasive argument against continued research, developments and applications in genetics has been made. None of the arguments against genetic engineering have been able to touch any of our cultural, ethical or religious values in a way that is relevant to a critical evaluation of genetic engineering. Because of this, business has continued fairly much as usual and it appears likely to do so for the foreseeable future.

B. Models of Scientific Research

Another problem in the relation of science and society is a change in the way of doing science or in the model of scientific research.[3] Traditionally, the purpose of science and other related disciplines has been to discover the truth of nature. The scientific quest was to understand what made things and organisms work. Research focused on discovering structures and stating general laws. Once this was done, the primary task of science was finished.

This model is changing, primarily because of discoveries in genetics as well as the application of a variety of other scientific principles, especially that of nuclear energy. The new reality is that not only can we know the truths of nature, but we can also change nature to suit our needs—or wants.

The discovery of the structure of DNA by Watson and Crick in the early '50's set in motion a chain of events that has led to the reconstruction of the DNA molecule. It is now possible, as mentioned earlier, to reconstruct various molecules and to make them perform in new ways. It is equally possible to build a new species out of previously existing species and in this way intervene directly in the evolutionary process. Stating the basic scientific laws that regulate the workings of nature has given way to the capacity to intervene into the very heart of nature by changing the genetic code to make a new product.

C. Responsibilities of Scientists

Such mighty powers have even more profound social implications. Although the caricature of the scientist sitting in the research laboratory, unencumbered by any of the cares of the workaday world, is manifestly inaccurate, nonetheless many scientists directed their primary efforts to basic research with little worry of potential applications or of long-term implications of what they were doing. It was assumed that they were apolitical and only provided information which society would then determine how and when to use.

Given their new powers and status within society, scientists face new responsibilities. These include an examination of their responsibilities for the application of their discoveries and an evaluation of the new role scientists are playing as policy advocates.

Daniel Callahan has suggested four general propositions which are helpful in an initial re-evaluation of the responsibilities of the scientist.

1. Individuals and groups are ordinarily responsible only for the consequences of those actions if, through negligence, they fail to take into account such consequences.
2. Individuals and groups cannot be held responsible for those actions the consequences of which are totally unknown. However, if they voluntarily undertake such acts, they may be held responsible for the consequences unless there were serious reasons for undertaking the action in the first place. One cannot, without serious reason, just "play around" in the unknown while simultaneously disclaiming responsiblity for the results.
3. When others may be affected by our actions, they ordinarily have a right to demand that their wishes and values be respected. This is particularly the case when those actions may result in harm to them.
4. Individual scientists and scientific groups are subject to the same norms of ethical responsibility as those of all other individuals or groups in society. They have neither more responsibility for their actions nor less; there is not a special ethic of responsibility applying to scientists that does not apply to others.[4]

Callahan then supplements these general propositions with two principles that he derives from our past experience. The first of these is the historical principle. This principle suggests that we know, in ways that earlier generations did not, that the search for knowledge can bring about harmful consequences and that it is possible to trace back the causal sequence. On this basis, Callahan argues that we should evaluate more carefully research that can potentially set in motion causal chains, of which some outcomes might conceivably be harmful.

This historical principle is supplemented with the imagination principle which suggests that a scientist might well assume that since unintended harmful consequences have happened with other forms of research, the same thing could happen with this particular project. Therefore, it is incumbent on the scientist to try to imagine or envision possibilities that may arise in the application of the particular project so that such applications can be evaluated. Such a framework, while not totally fail safe, provides at least a context in which critical evaluation can occur before a project is started or when it appears that there may be harmful outcomes from a particular application of knowledge from a project.

The on-going debate over the safety of recombinant DNA research and the continuing debate over the safety of nuclear power has brought forth a new model of a scientist: the scientist as advocate. In these debates, we see different scientists hurling technical—as well as personal—accusations at one another. Experts in a field can and do disagree on both the facts and the interpretation of those facts. We have typically assumed that such differences have occurred in the past. But we have not seen them in the full light of various media and city councils. These valid disagreements, however, allow scientists to be perceived by both the public and their peers as advocates for a particular position or cause. Scientists realized that they were assuming a new role and were often uncomfortable with it. But they also realized that the issues they were discussing were socially important enough to justify such a shift in role.

Of course, the model of the scientist as advocate presents an interesting problem about the relationship between facts and values. Oftentimes the facts may be reasonably clear and apparent. But the framework that scientists use to evaluate these facts may be different because of personal and/or social reasons. For example, one scientist may accept different risks than another and this will color how a scientist evaluates a particular problem. Thus, it is important for the scientist, when he or she assumes or falls into the role of advocate, to recognize as clearly as possible the personal, scientific, social, political, and cultural sources of his or her viewpoint so that the advocacy can be conducted on as explicit a basis as possible.

Our consideration of these three issues has shown that we are at a critical point of transition. We know that the way science is done is different. We have new capacities. Scientists are functioning in a new way. While still perceived as experts, their expertise is no longer accepted as value free. We assume that scientists are advocating particular applications as well as explaining them. New responsibilities arise from this shift and these have not been fully understood, explained, or accepted by either scientists or the public.

Ours is clearly an age of transition. We know our past but our future is uncertain. We must make decisions, choose particular paths, make specific applications. Typically, we have relied on the knowledge provided by experts. We have seen that a change is occurring here. What about our knowledge? Is it undergoing the same change as science? This is the problem for the following chapter.

3
The Problem of Knowledge

A. General Problems

We have all known the pains and anguish that come from knowing too little, especially about how to cure a particular disease that is killing an individual. Lack of knowledge makes us helpless in resolving a particular problem or in developing strategies to provide for different contingencies. Limited knowledge makes individuals hold on much too firmly to that which is known for fear that if that is questioned the foundations will be shaken and the universe will collapse. Limited knowledge can lead to a repression of knowledge and we all know the terror that comes when the pursuit of knowledge is repressed.

In our day we have, first, to face the problem of too much knowledge. The information explosion has caused facts, as well as the journals and books that report these facts, to proliferate. Even within a very narrowly defined area of knowledge, a responsible professional can barely keep track of new developments. Another problem of the information explosion relates to the qualitative dimension of that information. We are being forced to ask questions for which we really have neither a good answer nor a sense of how to go about even approaching the question. Our traditional sources of values are being strained to their limits by our technical capacities that follow from our gains in knowledge.

Second, a variety of disciplines including genetics, psychology, psychiatry, philosophy, and theology have all coalesced and are raising questions about the definition of the human person. Until recently most of us were reasonably satisfied that we had a workable sense of who we were and what we were about. New discoveries and insights into the full range of possible meanings of the human have given us a new burden in that we are no longer quite so sure of who we are and what we are to be about. The culture in which we live certainly reinforces this uncertainty. But it also perversely casts us further adrift by providing neither a common basis nor a set of values as a foundation on which to construct a new answer to the question of what is a human being.

B. Should Knowledge Be Restricted?

In addition to these problems, another major issue is being raised: discussions of the possibility of limiting or restricting research. Freedom of speech and freedom of thought are, of course, two of the most cherished values of our American culture. Any suggestion that knowledge be restrained faces a most difficult challenge. In fact, the presumption is that any restriction of knowledge or thought is almost inherently evil. Yet if one keeps in mind the shift in the model of nature from one of discovering the truth to that of changing nature, the argument about restricting knowledge may also shift a little. The knowledge in question here is knowledge of application or the implications of a technology, not the underlying creative thinking that led to the development of the technology.

Few people argue that scientists should be restrained from thinking through a particular problem or speculating on a new theory. The traditional argument for this is well stated by scientist Key Dismukes:

A major factor in advancing scientific understanding and correcting error is the opportunity of critics to challenge prevailing views and, if they can, adduce convincing evidence, to modify an existing consensus. This aspect of science is more than a convenient and useful tradition. It is as essential to the operation of science as freedom of speech is to the maintenance of democracy.[1]

The critical problem arises, however, when such knowledge is translated into action. Here the lines are not so cleanly drawn. There is the tradition in American law that, for example, religious freedom is a freedom to believe whatever one wants. However, one typically does not have a similar freedom to act upon those beliefs. Additionally, some restrictions on actions are already in place, such as regulations concerning the participation of human subjects in biomedical and behavioral research and the regulations concerning the recombinant DNA technology.

One background issue in this discussion is the value of progress within our society. In many ways, the knowledge explosion is a direct result of our valuing progress. The unconscious or uncritical assumption is that progress is in and of itself valuable and, therefore, must be pursued. To achieve this goal, research and development are necessary and have a high priority. But it is legitimate, at least, to question the value of progress and its role in our society, even though General Electric may continue to argue that it is our most important product. If progress is not morally necessary, and perhaps even optional, then it may be the case that much of what we perceive to be necessary may be interesting but superfluous.

This is not a direct argument against progress itself; it is a suggestion that mindless developments with an exponential generation of data may be inappropriate at this time. This is especially so in a time of diminishing resources, e.g., money and energy. At a time when the total budget to be spent on scientific research and development is diminishing, it may be appropriate to target certain areas of research as having priority because of their social necessity and value. In this light, limitations on research and knowledge could come about, not because of inherent distrust of the knowledge to be gained or problems or concern about some of the applications, but rather because of the acceptance of a new system of social values and/or priorities. Greater care in selecting research projects to fund would of necessity limit knowledge both quantitatively and qualitatively.

Yet, as Daniel Callahan indicates, our society at present values both basic and applied scientific research.[2] Because of this cultural value, he argues that the burden of proof must lie with those who are opposed to the particular research. He makes two exceptions to this basic rule. The first applies to a case where serious potential harm to the general public can be calculated with a degree of probability greater than 0. When that is the case, those who wish to pursue the research must submit the issue to public discussion and judgment. A second exception arises when there is a high probability that the research could cause a harm that would be of such magnitude as to pose serious threats to human welfare. In that instance, Callahan argues that the research should not go forward at all, even if it would be supported by the public.

Such an orientation, building as it does on the current cultural status quo, is persuasive in its argument that the burden of proof rests with those who are opposed to research. Yet it does allow for the possibility of restricting research when there is a probability of harm and of prohibiting research when there would be serious harm to human welfare. Although not foolproof and containing several ambiguities, such a position allows the discussion of this critical issue within our contemporary social framework.

David Smith, of the Religious Studies Department of Indiana University, suggests several reasons for restricting freedom of inquiry.[3] The first of these argues that knowledge may be immoral in its use and, therefore, may be restricted. This position would argue that the right to know must be less than absolute because some knowledge can end up doing more harm than good. While recognizing the problems and limitations with this orientation, Smith argues that it is important to think through what consequences the use of knowledge will have.

He suggests that knowledge that is either obtained or disseminated in an immoral way should be restricted. For example, knowledge obtained at the expense of violating a person's integrity or privacy is immoral and such attempts to gain this type of knowledge should be restricted.

Third, knowledge which is disseminated in a way that destroys just cultural institutions or practices is immoral and should be restrained because it threatens the very fabric and basis of our communal life.

Finally, Smith suggests that knowledge which can be destructive of us as persons could be restricted. Some knowledge could shatter a person's world view and, Smith argues, perhaps such knowledge should not be communicated to that person. Similarly, he also suggests that premature communication of scientific theories could be immoral because they are untimely and, therefore, may also be harmful to a person's self-understanding.

The basis of Smith's argument is his perception that knowledge is social and must be evaluated in a social context, not in an exclusively individualistic framework. He also argues that a scholar or scientist has some responsibility for the repercussions of his or her publication of the research results. Therefore, each should evaluate what is being said, when it is being said, and why it is being said before the results are actually published. Thus, Smith concludes by arguing for self-disciplined hesitation and professional accountability rather than censorship or repression.

Again, our progress in knowledge and application of that knowledge has brought us to a moral quandary: if we know too little, we can't help people as we would like. If we know too much, we are paralyzed by our power. We are also aware that scientific discoveries can have profound consequences for the world view of various individuals.

Do we limit our knowledge? Do we shut down certain areas of research? Do we pursue knowledge, but limit the understanding only to experts? Do we wait for natural events—depletion of natural resources or federal deficits—to solve our priority problems for us? No matter what we do, our decisions will have a significant impact on those who come after us. Let us now examine the moral stake our descendants have in this debate about knowledge.

4
Our Descendants and Their Future

Concern for progeny has been a major issue for almost all humans. Whether this be a desire to have children, to perpetuate one's name, or to leave the world a better place, concern for future generations has played an important role in shaping present debates. We have just seen how decisions about whether or not to restrict knowledge will affect our descendants. How much more impact on future generations will the capacity to intervene directly into one's genetic structure have?

A. Stewardship or Co-Creation: Which Ethic for the Future?

The traditional ethical model in which personal responsibility was exercised in relation to the earth and one's descendants was that of stewardship. This doctrine takes its point of departure from the creation narrative in the Book of Genesis. As this doctrine was developed through the centuries, it was assumed that this stewardship was exercised by observing the limits inherent in the orders of nature and society. The order of society was based on the order of nature and that order was established by God at creation. Thus both orders were divinely sanctioned and set the boundaries for acceptable behavior.

In the light of the theory of evolution and advances in the science of genetics, some suggest that the concept of co-creation may be a more proper description of human responsibility. Interestingly enough, Robert Francoeur, a biologist and theologian, locates this perspective in the same biblical narrative.

> But it seems to me also that in our panic we have deliberately avoided one of the most basic premises of our Judeo-Christian tradition. We have always said, often without real belief, that we were and are created by God in his own image and likeness. "Let us make him our image, after our likeness" logically means that man is by nature a creator or at least a co-creator in a very real, awesome manner. Not a mere collaborator, nor administrator, nor caretaker. By divine command we are creators. Why, then, should we be shocked today to learn that we can now or soon will be able to create the man of the future? Why should we be horrified and denounce the scientist or physicist for daring to "play God"? Is it because we have forgotten the Semitic (biblical) conception of creation as God's on-going collaboration with man? Creation is our God-given role, and our task is the ongoing creation of the yet unfinished, still evolving nature of man.[1]

This orientation, while containing some overtones of a Promethean presumptuousness, suggests that humans now have the ability to enter into the process of evolution, to shape it, to direct it, and to redesign it by developing new life forms. As Karl Rahner, the late German theologian, noted:

> He no longer simply takes stock of himself, but changes himself; he contents himself neither with steering by his own history merely the alteration of his sphere of existence nor with the mere actualization of those possibilities which have always offered themselves to man in his commerce with his fellowmen both in peace and in war. The subject is becoming its own most proper object; man is becoming his own creator.[2]

The model of co-creator assumes that nature is dynamic and changing and that the end of the process is open, but related to the absolute future of humans. Responsibility is exercised in this model by shaping and directing the evolutionary process according to values and criteria perceived to be appropriate in the light of goals that will promote human and social goods. Given this new stage in development as well as a new understanding of nature the model of co-creator seems at least as appropriate as that of steward.

B. The Person: Status, Qualities, and Expectations

Along with this debate on the model through which responsibility should be exercised, there is a continuing debate surrounding the understanding of personhood. The discussion centers on both indicators of humanhood as well as on qualities appropriate for human beings in the pursuit of their ends and goals.

Joseph Fletcher made the initial contribution to the former debate by suggesting a variety of indicators of humanhood which included criteria such as minimal intelligence, self-awareness, self-control, a sense of time in the past and future, a capacity to relate to others, concern for others, communication, control of one's own existence, curiosity, changeability, a balance between rationality and feeling, idiosyncracity and neocortical function.[3] In a later article, Fletcher singled out neocortical function as the essential trait, the key to humanity.[4] He justified this because of the role the neocortex plays in providing the biological sine qua non of all human activities.

There were a variety of responses to Fletcher's original criteria which were more or less happy with them, depending on the starting point. Nonetheless, in spite of the somewhat cavalier attitude with which they were proposed, Fletcher did provide a service by pointing to several problematic areas in defining a human being and stimulating debate on these problems. Even so, we must remember that even widely accepted criteria of indicators of humanhood may not provide the total basis for determining the value of a human being. One has to, for example, give an account of why specific biological functions are valued.

Another contribution toward the understanding of personhood in terms of desirable qualities is provided by Alasdair MacIntyre.[5] MacIntyre established his criteria not by setting minimal criteria by which one would be judged to be a person or not, but by arguing for qualities that would be desirable in designing one's descendants. These include: an ability to live with uncertainty, an understanding of one's past which provides a sense of identity, the ability to engage in non-manipulative relations, finding a vocation in one's work, accepting one's death, developing the virtue of hope, and a willingness to take up arms to defend one's way of life. These are very suggestive elements that are important for understanding who persons are and how they relate to others, to society, and to nature.

The first quality MacIntyre suggests, the ability to live with uncertainty, is a very important virtue, especially in the light of evolution. The assumption that the orders of nature and society were normative provided ethical, psychic and social stability. These orders were perceived to be rooted in the eternal plan of God for the universe. We accept today that societies are the product of histories, and not a direct translation of a metaphysical order. One of the realities with which we must make our peace is the fact that our world is changing and will continue to change as the result of human choice in relation to the physical and social environment. Therefore, the ability to live with a lack of certainty is a highly desirable personal quality.

MacIntyre very wisely, however, roots the quality of being able to face an uncertain future in an identity that comes from a strong sense of the past, one's place in a family, in a neighborhood and in a community. Knowledge of self and one's origins provides the strength needed to face an uncertain future.

Two other qualities are very important. The first is the need to find meaning through one's work. There is a twofold suggestion here. First, we need to find meaning in our lives and one appropriate way to do this is through the vocation that we have in the world. Second, there are some things that are worth doing and it is important that they be done regardless of their consequences.

The other important quality is the virtue of hope which is belief in a reality that transcends what is available as present evidence. The virtue of hope helps take us beyond a purely rational orientation to reality and provides us with a larger framework with which both to see and to evaluate what we might be about as we face our uncertain future.

C. Conclusions

Both of these orientations toward understanding the person suggest important issues. Fletcher, in his way, emphasizes the role of rationality and planning in defining human qualities. In his other writings, Fletcher has continued to emphasize the role of rationality by arguing that the more something conforms to rationality, the more human it is. For Fletcher, the use of genetic engineering, screening programs, amniocentesis and the like to insure the birth of a perfect child is more human because such processes are more rational. On the other hand, MacIntyre looks at broader qualities which appear to make persons more human. He suggests, by implication, a stance toward nature which presupposes the rational, but transcends it through an attitude of humility toward the future. The virtue of hope gives us a positive orientation to the vast processes that unfold before us.

Both of these models have their strengths and weaknesses. By pointing to significant dimensions of personal experience and a sense of the self, both suggest a variety of relationships toward nature that will be important in re-evaluating the two senses of responsibility toward the world described immediately above. In some respects, both of these orientations are departures from the traditional model of the person found in classical Western philosophy and theology with its emphasis on a static personal nature within a static social and natural world. But they are important because they suggest and allude to critical dimensions of the person that were not fully taken into account by that classical tradition. Thus they are extremely helpful in elucidating qualities necessary to cultivate as we begin to redefine our new place in a changing world.

Another important issue is the articulation of one's relationship to one's descendants. This question, of course, looms large on the horizon because of our growing perception and experience of the scarcity of resources. We have already left our descendants a damaged environment and a world depleted of many of its resources. This is not a very positive statement about ourselves and even less of one about our concern for others.

Certainly our descendants, whoever they may be, will have a number of interests similar to those that we have. The problem is trying to define the basis on which those interests should be respected, if at all. Although utilitarian and contract models of society may not provide totally satisfactory resolutions of the problem, both suggest that one should at least look to the future when calculating total utility or when trying to define how to act justly. Another approach would suggest that we know that our descendants will need certain basic goods and that they are entitled to these as a matter of human rights. This orientation argues for not harming future generations rather than promoting their well-being. Both of these approaches might suggest that we should leave our descendants at least as well off as we are, for in so doing we respect their interests and leave them the resources necessary for an adequate quality of life.

One's orientation toward this problem would also be affected by how I see myself in relation to other human beings. If, for example, I see myself as a solitary individual with few links to my neighbors and my community, then the whole question of responsibility to others has a less significant place in my ethical calculations. If however, I see myself as part of a community which has come from other communities and which will produce future communities, then it is more likely that I will be concerned with the environment that I hand on to my immediate descendants. These moral connections form the basis for evaluating my actions in the light of my needs as well as those of my descendants.

Another framework for analyzing this problem comes from one's orientation toward the end of the world. If one adopts a more apocalyptic viewpoint, then the question of future generations becomes somewhat less critical because when the end comes, it will come quickly and reality will cease to have significance. The apocalyptic orientation suggests that the world and nature may not be as teleological or goal directed as we would hope. While all of us may have goals and aspirations that we wish to see fulfilled, ultimately the world ends and we with it. On the other hand, if one has an eschatological viewpoint which sees the future as the source of goals and values, then one can build toward a reality that will come to fruition. In this framework, it is important to build a world for one's descendants that can be lived in and can be a continuing source of hope.

If one views the future through the apocalyptic lens, the question of the future is not that important because the end of the world is the end of significance. In the eschatological framework, however, what goes on within history and culture is important and stands in relation to the future that will, eventually, be reached. In this framework, the relationship to one's descendants is important and must be evaluated much more carefully.

This chapter raised the critical issues of images of personhood and responsibility to the future. How we evaluate persons and how we exercise responsibility will reveal much about us, our values, and what we think of the future. And this also sets the ethical agenda for the rest of this book.

5
Biology and Behavior

Descriptions of how our biological structures and our human behavior interact have produced a complex and fascinating history. Theories of determinism and free will set the broad parameters for the debate. This debate has then influenced how we understand traditional ethical concepts such as responsibility and altruism. The new discipline of sociobiology and a more sophisticated understanding and knowledge of genetics have added a new and controversial element to this traditional debate.

A. Altruism: Noble Behavior or Our Genetic Program?

Let's begin by considering the example of altruistic love, expressed as either giving one's life for another or in being one's brother's or sister's keeper. This concept has been one of the major pillars of the Western ethical tradition. But a suggestion of sociobiology is that perhaps such an exercise of altruism is not entirely voluntary and, therefore, may be a behavior for which the individual is not totally responsible. Rather, such behavior may be our genetic program. Thus our altruism is simply a mechanism of biological survival rather than an important moral virtue.

The British biologist B. Haldane concretized the genetic structure of such an understanding of altruism when he said he would lay down his life for two brothers or eight cousins. It takes that many of each group to achieve a genetic identity to his. Only this biological equality would make his own sacrifice genetically acceptable, i.e., insure the same number of similar genes remaining in the gene pool.

The basic implication of such a posture is that one is altruistic toward those who are genetically similar because even if I do not benefit directly, I benefit those who have similar genes. Therefore, from a biological perspective, the survival of any particular individual is not important. The survival of similar genes so they can replicate themselves is what is critical.

In the framework of E.O. Wilson, one of the contemporary founders of sociobiology, egoistic behavior is behavior which guarantees that the genes which cause copies of themselves will survive. Altruistic behavior insures that copies of genes that an organism contains will survive, although they may be contained in another organism. These biological definitions of egoism and altruism are then used interchangeably with the ethical concepts of selfish and unselfish.

The problem is that a direct translation of these terms appears to be rather difficult. On the one hand, behaviors that we experience as selfish or unselfish are usually conscious and the result of an evaluation of consequences. A strategy for genetic replication is typically unconscious and, therefore, not under our control. Also the way in which the words egoism and altruism are used refers primarily to actions which affect the gene pool. One could infer that actions which have no significant impact on the gene pool must be neither egoistic nor altruistic, or, in value terms, selfish or unselfish. That, however, does not correspond to our experience. Therefore, we must be wary of such an easy and uncritical translation of biological categories into ethical categories.[1]

Even though Wilson may not have the precise translation of genetic terms into ethical terms, Arthur Caplan, a philosopher at the Hastings Center, argues that there is a point at which such verifications of biological behavior would be relevant to ethical theory. He illustrates this by a discussion of psychological egoism and ethical egoism.

The theory of psychological egoism is a factual theory about human motivation which claims that persons always try to act in their own best interest. If such a factual theory were true, Caplan argues that the only reasonable basis for ethical behavior would be a theory of ethical egoism which postulates that morality is simply an expression of self-interest. This would mean that the only acceptable and meaningful ethical principle would be: act always to promote your own individual good as much as possible. A significant part of the argument for this position would be determining both that genes actually cause specific behaviors and that they are the sole causal agent for them. In addition to the empirical data needed to prove this position, one would also have to accept a great deal of reductionistic theory which has its own theoretical problems.[2]

B. Freedom and Determinism

Sociobiology is also involved in discussions of freedom and moral responsibility as mentioned above. First we must clarify the concepts of freedom and determinism. If freedom means a radical freedom in which a person is bound by no constraints whatsoever, then sociobiology would join forces with traditional philosophy and theology to argue against such a position. On the other hand, if one posits a more modest account of freedom, then sociobiology may not be able to make its case as strongly as its proponents hope.

Discussions of freedom have to be related to discussions of determinism. Two extremes can be posed. Hard determinism holds a theory of universal causation which argues that for every event and effect, there is a cause. Freedom, by definition, is incompatible with this state of affairs. Sociobiology would argue, accordingly, that each of us is genetically determined and, therefore, subject to irresistible compulsions and coercions which cause our actions.

Soft determinism also holds a theory of universal causation. However, it differs from hard determinism in that it argues that some causes originate with the individual. Biologically, one could argue that each of us has a set of predispositions, but that such predispositions do not determine any one particular action. This allows a modified understanding of freedom.

Hard determinism eliminates both freedom and moral responsibility. Assignment of moral responsibility requires some degree of personal causality in acting. If one cannot determine or discover such causality, then one must argue that the individual human is not morally responsible.

Human behavior does not seem reducible to a set of biological coordinates. The reductionism necessary to establish this is contrary to our conscious experience. Similarly, it cannot account for our experience of freedom of choice and the common assumption of responsibility for one's actions. Soft determinism is compatible with a theory of human freedom in that it argues for human freedom and causal efficacy within a genetic and environmental context. A theory of human responsibility is set forth in a context which assumes a variety of causal agents operating at different but related levels.

C. Social Policy Implications

The belief that all persons are created equal is a cornerstone of the American way of life. Yet over the past several decades many allegations have been made about genetic differences between people that challenge this claim of equality. Some have suggested, for example, that intelligence is a function of one's racial group or that aggressive behavior is related to the presence of an extra Y chromosome. These genetic differences would then require, it is argued, different social treatment of such individuals.

A particular social policy does not necessarily follow from a set of empirical facts. That an individual may belong to a group with less genetic potential does not, by that account, mandate a particular policy. Thus, for example, one could, on the basis of this set of facts, just as easily set a policy of protection as well as "benign neglect." What is relevant for setting policy is the ethical and political framework in which these facts are evaluated.

Such established facts about genetic potential should be taken into account when setting policy. Otherwise, an unrealistic policy or goal may be set which would only frustrate the individuals in question and, perhaps, set up a self-fulfilling prophecy. We should not legislate or mandate that which we are not capable of—biologically or psychologically. Thus, a just policy would require attending to the genetic potential of individuals, but in such a way that these individuals are neither discriminated against nor deprived of sharing in the basic goods of the society.

Facts need to be taken into account in setting policy. However social and cultural values mediate those facts as they are incorporated into policy. Genetics does not provide a totally adequate basis for such a social evaluation because it provides only one element of what is necessary for an adequate social policy—potential. Genetics does not specify how to evaluate morally that potential.

D. Conclusion

Sociobiology and new discoveries in genetics suggest that there are problems for those who think equality means sameness. People are biologically different and, consequently, have different potentials. However, the fact that people are unequal with respect to potential does not mean that they are unequal with respect to moral status. A better understanding of the biological basis of behavior would keep us from making unreasonable demands on people as well as help us establish a more adequate understanding of human responsiblity. In this way, sociobiology can make a significant contribution to ethical theory and social policy.

6
Points of Intersection:
Nature, Health, and Ethics

Before we proceed into areas of specific application of areas of genetic engineering, we need to look at two other areas—definitions of nature and health and their implications for ethics. This chapter will set a context and provide a further frame of reference for our consequent discussions.

A. Definitions of Nature

1. Nature as a Limit

In discussing our moral responsibility to our descendants in Chapter 4, I referred to the concept of nature as a limit. This understanding is based on traditional Aristotelian philosophy which postulates an order in nature which reflects the pre-existent plan in the mind of the Creator. The order of society reflects this natural order and thus incorporates the plan of God into the fabric of daily life. The philosophy of natural law, derived from this model of reality, derives moral norms from the order of nature. Because this order of nature reflects the plan of God, it is normative and constitutes the limits of society and human action.

The traditional Roman Catholic argument against artificial

contraception is based on this model of nature as a limit. In simplified form, the argument states that what nature—and thus the plan of God—united, no one can separate. Thus the unity of sexual intercourse and procreation is inherent in the order of nature and prohibits any artificial separation of them.

Biological structures reflect metaphysical principles and thus constitute the limits of personal and social behavior. Such a moral tradition tends to be conservative and would approach interventions into nature with caution, if not suspicion.

2. *Nature as a Model*

This orientation, similar to the natural law perspective above, argues that we can replicate some occurrences in nature. Thus in this framework it might be possible to replicate certain pre-moral evils that occur in nature as long as one has a proportionate reason for so doing.

For example, a fairly high number of zygotes are lost during the implantation process. A number of these seem to miscarry because of structural anomalies, hormonal imbalances within the uterus, or simply a failure of many different complex systems to interact correctly. Is it moral to replicate such embryo loss in the laboratory in an attempt to fertilize eggs in vitro? Benefits to future embryos and fetuses as well as to women unable to conceive in vivo could then justify many research protocols. One would be replicating a natural phenomenon in the laboratory and justifying such embryonic loss on the basis of the benefits to be achieved.

Such an argument is put forward by Richard McCormick, a Jesuit priest and a senior Catholic moral theologian.

> It is not a violation of the right to life of the zygote if it is spontaneously lost in normal sexual relations. Why is it any more so when this loss occurs as the result of an attempt to achieve a pregnancy artificially.[1]

Karl Rahner, basing his orientation on the doubtfulness of the personhood of the fertilized ovum, suggests that zygotes could be used as subjects of experimentation.

> But it would be conceivable that, given a serious positive doubt about the human quality of the experimental material, the reasons in favor of experimenting might carry more weight, considered rationally, than the uncertain rights of a human being whose very existence is in doubt.[2]

This orientation is countered by Leon Kass, a physician at the University of Chicago Medical School.

> Parenthetically, we should note that the natural occurrence of embryo and fetal loss and wastage does not necessarily or automatically justify all deliberate, humanly caused destruction of fetal life. For example, the natural loss of embryos in early pregnancy cannot in itself be a warrant for deliberately aborting them or for invasively experimenting on them in vitro any more than still births could be a justification for newborn infanticide. There are many things that happen naturally that we ought not to do deliberately. It is curious how the same people who deny the relevance of nature as a guide for re-evaluating human interventions into human generation, and who deny that the term "unnatural" carries any ethical weight, will themselves appeal to "nature's way" when it suits their purposes. Still, in this present matter, the closeness to natural procreation—the goal is the same, the embryonic loss is unavoidable and not desired, and the amount of loss is similar—leads me to believe that we do no more intentional or unjustified harm in the one case than in the other, and practice no disrespect.[3]

3. Nature as Evolving

The previous two models of nature assume that it is relatively static. This model of nature sees nature as continually evolving. While not suggesting that there is no stability to nature or that there are no laws of nature, the model suggests that such stability may not be as normative as in other models and that, consequently, more interventions may be justified. This view would also tend to see change or development as normative rather than exceptional.

Within this model the concept of history takes on an importance lacking in the other two models. History is linear or teleological, rather than cyclic or episodic. History has a future and that future draws history forward. In Christianity, this orientation presents a dichotomy: the ultimate future transcends persons and their efforts at self-creation, but, also through their interventions in history and nature, persons open themselves to this future and help to achieve it. Rahner says:

> This human self-creation will develop the concrete form of human openness which leads to the absolute future that comes from God. But it is never capable by itself of bringing about this absolute future. Christianity, precisely because it is the religion of the absolute future, must simultaneously send man out to his duties in the world.[4]

4. Conclusions

We have an increasing capacity to intervene into ourselves as a result of developments in genetics, psychology, and many of the behavioral sciences. Such capacities are in tension with the first two models we discussed but are assumed by the third.

We no longer experience the past as normative, and the future assumes a greater role in defining efforts at self-creation. Such capacities situate us between a static understanding of our nature and the ability to shape evolution. MacIntrye's virtue of hope is necessary as we face our open future.

B. Health and Genetics

A thematic issue raised by genetic engineering and genetic screening programs is the question of what is health and what is a disease. In addition to the medical dimensions, these definitions are also important because these perspectives frequently help determine whether or not to intervene.

One orientation suggests physiological definitions of health and disease.[5] That is, health is functional normality. Such functions are determined by design, conformity to goals pursued by the organism, as well as the capacity to work out the design of the organism. Each organ or system of the body has a specific range of activities, and if there is deviation from that range, there is disease. If the disease is disabling, then it is an illness.

Paul Ramsey, a leading Protestant ethicist, incorporated this perspective into an ethical argument called a medical indications policy.[6] Decisions to treat or not to treat should be made primarily on the basis of physiological criteria. The base line for determining whether or not to treat is whether or not the treatment will benefit the patient. In addition to accepting a physiological definition of health, Ramsey also assumes that these determinations are made in a value free context.

A second orientation argues that in addition to a physiological component, definitions of health contain social and cultural dimensions. This is particularly true in the areas of psychology and psychiatry, but such value considerations influence how we evaluate genetic diseases. A broken arm, for example, presents little difficulty in diagnosis and prognosis. But when one is diagnosing mental illness, many interpretations are possible, depending as they do on one's theory of mental illness and one's social, ethical and cultural values. While there certainly is mental illness, there is also an ideological, theoretical, and value dimension to our perception of those individuals who fall outside what society perceives as normal or socially approved behavior.

These perceptions can affect an individual with a genetic disease in a variety of ways. Let us use Down's Syndrome as an example. There are clear physiological criteria by which one can diagnose the syndrome. There are established developmental markers by which one can gage the severity of the syndrome and make predications. However, in our culture, intelligence is highly valued and because of this the diagnosis and prognosis are perceived differently. In addition to having a syndrome, this individual is defined as impaired and may be socially disvalued because of diminished capacity for abstract reasoning.

Others have suggested that knowledge of an individual's genetic constitution may alter how that individual is raised. Take, for example, the XYY Syndrome. Some have argued that this is a possible predictor of violent or aggressive behavior. If parents were to learn of this, would this not alter how they would raise the boy? Should this boy be defined as ill or healthy?

Here is a genetic anomaly that has a clear physical marker, but an unclear social expression. How one perceives genetic anomalies, aggressive behavior, and the influence of genes on behavior will have a significant impact on how the individual is perceived and accepted by society. Thus interaction of values from various sources set a context in which a physical reality is evaluated.

Another problematic area is the often unclear distinction between a carrier of a genetic disease and one who is afflicted by the disease. A carrier is not afflicted by the disease. Genetic screening programs identify both types of individuals. But if the distinction between them is misunderstood or confused, then carriers could be prevented from receiving insurance policies or other health care benefits because of a wrong assumption that they are unhealthy. Not only may such individuals be unjustly deprived of a variety of benefits, they are unfairly labeled, and this can serve as a basis of discrimination.

C. Conclusions

The important point that I would stress as a consequence of our considerations of models of nature and definitions of health is the presence of a value framework that influences our most basic perceptions. We see and experience in a context. And whether this context is the cultural status quo, a particular ethical theory, or a definition of health, the values in this context influence what we experience and thus help shape our expectations.

What we have discussed so far in the book is the variety of ways in which different commentators have defined these contexts of perception and have identified various methods of interpretation. The remainder of the book deals with the application of these and other ideas to specific areas of genetics and genetic engineering.

7
Genetic Engineering:
An Introduction

The general ambivalence with which many people approach technology is reflected in two stories in the Book of Genesis. In the story of Noah (Gn 6—9) we see how God used a technology to save a faithful remnant while the rest of the world perished. Here, one can argue, technology is a means of salvation, and through the proper uses of technology, humans have a means to achieve important ends.

The story of the Tower of Babel (Gn 11) shows what happened when humans attempted to become the equal of God through the use of technology. Here the misuse of technology resulted in the destruction of human communality.

While one can raise questions of literal interpretations of these stories, nonetheless they are significant because they reveal attitudes toward technology, its purposes, applications, and consequences. Let us now examine several other themes as an introduction to several of the ethical problems in the vast area of genetic engineering.

A. The Quality of Human Genes

One major motivation for genetic engineering has been fear of a decline in the quality of the human gene pool which some think will result in a genetic apocalypse. In this scenario, deleterious or harmful genes would be passed on to our descendants whose quality of life would then be compromised. The biologist Theodosius Dobzhansky stated the issue this way:

> If we allow the weak and deformed to live and propagate their kind, we face the prospect of a genetic twilight; but if we let them suffer or die when we can save them, we face the certainty of a moral twilight.[1]

This short quotation presents several problems. First, who is the "we" to whom Dobzhansky refers? Are they scientists; are they physicians; are they politicians; are they parents? Second, whoever they are, what authority do they have and how did they obtain it? Third, why will elimination of variation from the gene pool bring only positive results? Fourth, how does this position square with the traditional Judeo-Christian tradition of concern for the weak and helpless?

From ethical and scientific perspectives, the issues raised by Dobzhansky are problematic. How we handle these issues politically, culturally, and ethically is also problematic. Thus while Dobzhansky raises important health problems, even more problematic is how we resolve these issues.

B. The Person and Nature

For the past several decades we humans have been celebrating our liberation from many of the tyrannies of nature. Developments in science and technology have allowed us to eliminate much needless suffering. Infant mortality is decreasing, our life span is expanding, many diseases have been eliminated, various organs are replaceable or can be compensated for, and increased understanding of the genetic code has opened up new potential for intervening in human beings.

But has our liberation from nature made us slaves to technology? Many people argue that there is a technological imperative at work. If we can do it, we should to it. Capacity confers justification. Such an imperative leads us to be uncritical in the face of what is being proposed.

The fact that options and interventions are new does not necessarily make them problematic. But neither does it make them morally indifferent. We may be putting ourselves in a situation which has fewer and fewer exits. We must do all we can to save lives, to provide children for those unable to have them "naturally," and to redesign animal forms. We intervene in nature so we can master nature. But we seldom ask what consequences this will have for us in the short and long run.

C. Purposes of Genetic Technologies

The ability to design and actualize specific goals for human, animal, and plant life is one of the most significant developments in genetic engineering. Our control over nature is clearly demonstrable and we are moving closer to the time when we can engineer the engineers.

Yet we tend to forget that humans control science. Science is not some independently acting agent. It does not choose what to pursue. Humans choose what goals to pursue through the application of science and technology. And it is this pursuit of goals that actualizes the value issues with which we are concerned.

At the heart of genetic engineering—as well as all of science and technology—stand very clear value choices that are made by humans who have certain priorities which are then actualized in a scientific and political context. On what basis are we sure of the desirability of our value choices? As we pursue specific ends and gain greater success in the application of our knowledge, we need continually to attend to the process of justification of our goals. Of concern also is the status of those who may disagree with those goals.

D. Cultural Dimensions

Americans have been particularly fascinated with and have a strong desire to achieve perfection and superiority. Advertising slogans tell us that progress is our most important product and that we will achieve better living through chemistry. The eugenics movement was concerned with improving or at least maintaining the status quo with human stock.

What does progress mean? What qualifies as better living? In a social context, what does superior mean and what social implications follow from that?

In typical discussions about improving the species, the talk usually turns to intelligence and the capacity for abstract reasoning. Seldom does one hear talk of trying to instill or develop the qualities of kindness, compassion, sympathy, or mercy.

Clearly, such discussions are rooted in the cultural status quo and reflect the prevailing wisdom in America. Yet, why is intelligence more important than compassion, a capacity for abstract reasoning more important than human decency? Obviously none of these are incompatible, yet traditional virtues governing interpersonal relations always seem to take second place.

E. Conclusions

Genetic engineering offers the promise of a substantive improvement in the quality of the life of future generations as well as an improvement in the quality of our environment, agricultural processes, and livestock development. All of these promises, improvements, and applications occur within a socio-political context which contains and reflects significant value dimensions. These need to be examined before we implement various scenarios in genetic engineering. Genetic engineering is not value free and the value or ethical dimension needs continual examination to raise and evaluate the desirability of the goals we propose.

8
Applying the New Genetics: Recombinant DNA Research

The process of recombining genetic material from one organism to another has made possible new and exciting developments with the area of genetics.[1] Techniques are available whereby the basic genetic material of an organism—deoxyribonucleic acid or DNA—can be cut up and recombined in new ways: putting genes from one bacterium into another and also transferring genes from viruses or animals into bacterium.

This technique has led to a greater understanding of the genetic code, the set of instructions whereby an individual develops himself or herself. It has also enabled the production of organic materials as diverse as insulin and a bacterium that eats oil. Existing patent laws are being strained by this technology because attempts are being made to patent life. Many companies based on these exciting technologies have recently been formed—and gone public—in the hope that their applications can produce significant breakthroughs in medicine, science, and industry.

The debate on the technology of rDNA has gone through various phases. As early as 1977, Willard Gaylin of the Hasting Center identified the rDNA debate as the Farrah Fawcett-Majors of scientific issues because one could hardly open professional or popular publications without encountering yet another article on the topic.[2] Although Ms. Fawcett has dropped the hyphen and Mr. Majors, articles still appear on the technology, but with less stridency of debate. More interestingly, these articles more frequently appear on the economics pages of journals. The debate has shifted from issues of safety and risk to the relation of industry and universities. Let us examine now several of the main themes of the debate and its current status.

A. Basic Issues

1. Safety

The first safety issue to be debated was the security of containment facilities. How secure was the laboratory at which the research would be carried out? Could the genetic material being used escape from the lab into the environment and cause harm? These questions led to developing regulations which established various levels of security within labs appropriate for different types of experiments. Special biodegradable research organisms were also developed. Although basically a technical issue, the social and environmental implications are profound.

The second issue is the safety and health of the scientists and lab technicians who work in these facilities. The bacterium typically used in this research, E. Coli, is one that has been well studied but is also one which lives in the human intestine. Whether E. Coli used in research could actually live in the human intestine after its rarified laboratory life is unknown. But there was fear that a genetically engineered form of it could escape, be ingested, and cause harm to the individual.

Again while primarily a technical issue, this concern led to regulations on appropriate lab techniques, the development of safe containment facilities, and the encouragement of proper laboratory habits.

The third issue was presented by the ecology. Since the use of rDNA can develop new life forms, many were concerned about their effects in or on the environment. Specifically, if a pathogenetic or carcinogenic material used in a research project were to escape into the environment, its effects were unpredictable. Many questioned the wisdom of creating new life forms to begin with. Finally there was concern about the environmental and evolutionary effect of the interspecies transfer of genetic information. These problems remain unresolved.

2. Conclusion

As indicated above, several of these issues are technical and have been dealt with on that level. Lab standards have been upgraded and secure facilities developed. Yet human frailty exists and the lab may not be the best place to test Murphy's Law.

Three basic safety issues need continual re-examination:

1. Will our legitimate concern with safety make such techniques and experiments so costly and difficult to perform that scientists will simply abandon this area of research?

If we insist on unreasonable safety standards, this may deprive us of certain benefits. Yet if we do not insist on standards that are safe enough, we may cause irreparable personal and environmental harm. Thus far the safety record has been excellent and has not impaired research. But as different techniques are developed and new experiments proposed, issues of technical safety may become a priority again.

2. Who will have the responsibility of deciding to accept whatever risks are involved in this research?

Typically the ones who do this are the scientists, the universities where the research is carried out, or an industry developing a particular process. The government is also involved through regulations established through the National Institutes of Health and by deciding which projects are to be funded. How the public can be involved, other than through the electoral process, is unclear at the present time. Thus experts retain their traditional power of decision-making.

3. How will decisions be made to develop new forms of research?

This has to do with evaluating the level of risk to the individual, the community, and the environment. The basic problem is that appropriate risk/benefit methods of analysis have yet to be developed. The enormity of the environment and evolutionary impacts compounds the issues dramatically.

B. Resource Allocation

We live at a time of dwindling resources—financial as well as material—and this issue is exceptionally critical. There are two fundamental levels of this problem: which programs to fund and at what level of funding. This double allocation problem raises several problems in terms of the relation between federal, state, and local community priorities, the relation between basic and applied research, and determining what will best serve our short and long term needs.

We are used to cost-benefit analysis as a means of allocating resources but the method falters on a national level of funding research. No one proposal is so clear in its benefits and consequences that it is the obvious choice for funding. Additionally, cost-benefit methods may not help us resolve the tension involved in choosing between basic and applied research.

Coordination of national priorities in the area of science and technology, the establishment of national and state priorities in terms of major needs, and the development of as precise a research proposal as possible are issues of highest urgency.

C. The Decision Making Process

1. Who Decides?

The previous discussion of resource allocation raises one aspect of decision making. Another is who the decision maker ought to be. Many argue that since rDNA is a scientific problem and procedure, scientists should be the decision makers. Others argue, however, that since many of the problems associated with rDNA are funding and priority problems, the people via the federal government should be the decision makers. Since the money is public, the decision belongs to the public.

Because of their expertise, scientists must clearly be involved in the decision making process. Yet decisions about proceeding with a particular program in rDNA involve setting both national and scientific priorities. Thus these decisions are a blend of scientific expertise and national priorities.

In making these decisions, two problems are particularly to be avoided. First is the fallacy of the generalization of expertise which assumes that, since a person is an expert in one area, he or she is also an expert in another. The second is the use of technical criteria as the sole basis of decision making. The assumption here is that if one has all of the technical data available, one has all one needs to know to make appropriate decisions. Are the technical criteria inclusive enough and are they, in fact, purely technical?

2. On What Basis Are Decisions Made?

As just noted, one basis for decision-making can be expertise. This is a mandate for developing technology assessments and environmental impact statements. Yet, as we know, experts often disagree and critical choices are made on the basis of unstated or unconscious value judgments, political pressure, or unclearly established priorities.

A second basis for decision making is authority. Should the authority of scientists go beyond their well-earned and well-recognized prestige within their field? Clearly scientists have authority within their field. Does this authority extend to the political process? Unless scientists are elected officials, they have no political authority. And even in the case in which a scientist is also an elected official, the authority of that person comes from the political process, not scientific expertise.

Public policy decision-making is fraught with value issues. To make informed and responsible decisions, we need to surface those values so they can be debated. Then one can determine how a scientific proposal relates to those values.

D. The Benefits of rDNA Research

1. The Promises

The recombinant process has been hailed as the cure for cancer, most human diseases, a means of increased food production, a way to clean up the environment and the way to significantly improve our livestock. Others, of course, claim that these projected benefits are all scientific hyperbole, gross exaggerations, and are inflated to increase funding from both government and industrial sources.

While such claims are not new within the history of science, one must take care to distinguish general benefits from specific benefits. Better health as a payoff of this technology may not be enough of a benefit to justify its costs. This may be particularly true if we can achieve many of these goals in ways that are not as costly: use of seat belts, reduction of pollutants on the environment, raising the drinking age, and decreasing or stopping the use of drugs and smoking. This is a carryover from the allocation issue. If funds for research are tight, then we must be all the more certain about the benefits to be derived.

2. *Actual and Potential Benefits*

We must also distinguish between actual benefits and potential benefits and relate that to the time frame in which either might be available. Discipline is required to separate facts from wishes. Research programs must be funded on the basis of what they can realistically deliver within an appropriate time frame. Otherwise they may be jeopardizing future funding in other important areas. Our hopes that benefits may come from a particular research program are not enough to justify that particular program, especially when funds are restricted.

3. *The Cost Factor*

We also need to know how much these actual benefits will cost to determine whether this outcome is worth this price. Are there alternative cost-effective measures to achieve the same benefit? Given a reasonable equality of costs, which program has the most potential to deliver on its promises? We are not seeking mathematical certainty about the actual costs, but rather are examining critically different programs to see whether these are the only alternatives available or whether other means might deliver the same kind of results.

E. The Responsibility of Scientists

1. *Personal Responsibility*

The scientist is obviously the front line of defense because the scientist develops new ideas and carries out the research to test them. Like everyone else, scientists are caught in a variety of interests, many of which squeeze him or her with a particularly fierce force. The scientist has a professional commitment to the enterprise of science; a professioinal code of ethics to which he or she is responsible; political loyalties to particular policies that he or she thinks are most worthy of implementation; personal loyalties, typically to his or her own profession, to the team of which he or she is a member, as well as to the intellectual challenge of discovering the truths of nature. All of these are worked out in a political, professional, and personal context and shape how the scientist begins to articulate his or her particular stance toward new technologies.

2. Professional Responsibilities

What can we ask of the scientist, in light of this, in terms of professional responsibility? First, the scientist is a citizen and has the same civic responsibilities as any other citizen in debating priorities, resource allocation, and funding decisions of the federal government. Second, the scientist is a professional and is bound by the professional codes of ethics of organizations of which he or she is a member. Third, the scientist has a level of professional expertise that brings its own responsibility, including responsibility for assessing the consequences of particular applications of a technology insofar as this is possible. Scientists could also consider how specific forms of research might relate to social needs. Finally, the scientist should be aware of the social, political, and cultural context in which science is done and how this and the scientist's own values merge to form a particular orientation toward social responsibility.

3. Summary

The scientist as a responsible professional attempts to articulate these different dimensions of one's professional, civic, and personal life into a coherent whole. Thus in terms of professional responsibility scientists are no better or worse off than any other professional. They try to make the best of a very complex situation and, if they are doing their best to be responsible to themselves, their profession, and their society, there is not much more that one can ask.

While one may disagree with a particular decision that a scientist comes to, one has to take into account the various factors that shaped that decision and must assume that the scientist, like all other people, is trying to put together an appropriate blend of values and expertise to use in decision making. Like all of the rest of us, scientists need to broaden their vision. But they may not be segregated out for special treatment simply because they are scientists. As scientists they have the basic moral responsibility to think through the implications of their research and to articulate the value issues that are involved in this.

F. Changing Nature

Another thematic issue in recombinant DNA is changing nature. This particular technology is clearly one more step on the way to redesigning nature and exercising a powerful amount of control over it. This power frightens many people because they think we may lose control and will be irreparably harmed. On the other hand this power is extremely exciting because of the tremendous benefits that can be derived from redesigning, changing, or correcting nature as it is. A cluster of thematically related questions present themselves under the rubric of changing nature.

1. Are There Limits?

Are there any limits at all to what we can do and, if so, what are they? Typically our limits have been set by our knowledge of nature and our means of intervention. We also assumed, philosophically and religiously, that the universe was static and could not be changed.

Our awareness of the historicity of the universe and its evolving nature have allowed us to relate to nature in a new way. Our understanding of the genetic code and our capacity to take apart and recombine different elements of genetic material raise the question of limits in a new and critical way.

2. *Responsibilities to Future Generations*

What might our responsibility to future generations be? Are we obligated to leave the world better off or can we leave it the way we inherited it? Many of the interventions we have already made, especially in the area of energy and pollution, have caused significant changes in the world and have left it a less desirable place for our descendants. Might our new technologies provide a means of rectifying some of these damages or perhaps provide future generations with some alternatives for coping with the problems that we have left them?

3. *What Is Human Nature?*

While we know that human beings have evolved, we also have a sense of the uniqueness of the human person in the world of nature. We know that human beings stand in a unique relationship to all other areas of nature. How far can we carry self-experimentation? How far can we push our alleged adaptability to new circumstances?

The new technologies make possible combinations of human and animal genetic material. What we would understand these combinatioins to be and what status such entities might have within our world is totally unclear. The question of human identity is an important one and it is raised very sharply by the development of these new technologies.

G. Universities and Industry: A New Relationship

The final thematic problem in the area of recombinant DNA is the development of new biotechnical industries and their relationship to universities and their faculties.[3] Faculty members have always been industrial consultants and this has typically not presented too much of a conflict of interest. Frequently the government was the major funder of research and there were clear guidelines. Also universities and scientists have worked out mechanisms to determine the entitlements of both the university and scientists. However with the new technologies, conflict of interest becomes very critical in terms of who owns the product that may be developed as a result of such new relationships.

When university research is funded by an industry, it is unclear who will have ownership of the final product and what role the scientist plays for each party. Scientists are also facing a conflict of interest insofar as they may be employees of both the university and a research corporation. And, since industry is typically interested in specific applications, the scientist may also be torn between pursuing basic research for which funding may be uncertain and working on scientific projects for which funding is forthcoming from industry. This may shift the direction in which science proceeds within our country.

These are new problems that are emerging within our country that, as yet, have no clear solutions. Different models of relationships between industry, universities, and medical centers are now being elucidated. No one model has presented itself as the definitive solution to the problem of the conflict of interest on the part of the researcher, the university, the industry, and the government. At stake is not only the traditional question of who developed a particular technology or process, but also the enormous profits to be derived from both holding the patent and licensing the manufacturer of a particular process or entity.

Thus there is a need to resolve this issue so that scientists will not be caught in an untenable conflict of interest nor will our country possibly experience a misplaced priority because of the concern over short term profits as opposed to a basic understanding of the mechanisms of nature.

9
Birth Technologies

A. The Technologies

A variety of technologies dealing with birth have developed over the past several decades, some greeted with much publicity and comment, others quietly making their way into practice and being accepted without much fanfare. Each is different, each presents its own ethical problems, but they all relate to the area of birth and child development and raise common concerns about the status of parenting and the fetus in our society.

1. Sperm Banks

Perhaps the oldest, most well established, and accepted form of a birth technology is the sperm bank. Sperm is frozen and reserved for use perhaps by an individual who had a vasectomy or had undergone chemotherapy or radiation treatment, but wanted to retain healthy sperm by which to father a child. Some simply donate to sperm banks so others can use this sperm to conceive a child. Typically these individuals would receive a fee for this. Commercial sperm banks have been in place for many years and are fairly well accepted means of obtaining sperm for a couple who is having an infertility problem.

2. Amniocentesis

A second established technology is amniocentesis. In this procedure fetal cells are withdrawn from the amniotic fluid that surrounds the fetus and are analyzed to discover whether or not there is a genetic abnormality. One can also establish the gender of the child. This technology has had a tremendous impact on the birth process in several ways. First, it allows many fetuses who may have been aborted to be saved because they do not have the disease for which they were at risk. Second, parents have a choice in terms of whether or not to continue the pregnancy of a fetus that has a genetic disease. Third, amniocentesis can be used for sex selections when there is the possibility of a sex linked genetic disease or when there is preference for one gender over another.

3. In Vitro Fertilization

A third technology that has drawn much attention is the process of in vitro fertilization. This technology involves the external conception of an individual and the subsequent implantation of the conceptus into the prepared uterus. Gestation then proceeds in the traditional fashion. This technology is particularly useful for women who are infertile by reason of structural problems of the reproductive system.

However, once having established the technology of external fertilization, one can clearly see that the egg and the sperm need not necessarily be from the couple trying to achieve a pregnancy. Thus neither the egg nor the sperm need be derived from either of the couple achieving the pregnancy, nor does the conceptus need to be reimplanted in the donor of the egg. This gives rise to the phenomenon of the surrogate mother whose uterus is prepared to receive a fetus conceived externally from genetic material from various individuals. Also this technology has been combined with others to allow embryos to be frozen so they can be reimplanted at a later stage. This procedure solves the problem of what to do with extra fertilized eggs. It also allows individuals the possibility to have an embryo in reserve for a later time. Also, logically, this embryo need not be returned to those individuals from whom the egg and sperm came.

4. *Embryo Transfer*

A fourth technology in this area is embryo transfer in which a woman achieves pregnancy in the traditional fashion or through artificial insemination. But after fertilization has occurred, the embryo is removed and reimplanted into the prepared uterus of another person. This avoids many of the technical problems associated with trying to achieve a pregnancy through artificial insemination or in vitro fertilization and reimplantation.

5. *Sperm Separation*

A final technology has to do with sex selection through the mechanical and/or chemical separation of the Y chromosomes from the X chromosomes in the sperm and using the appropriate concentration to help achieve the desired gender. What this concentration of chromosomes does is to increase the probability of obtaining the desired gender by perhaps 15 to 25 percent.

6. *Summary*

Most of these technologies are well in place. There are several in vitro fertilization laboratories set up throughout the country. Artificial insemination, whether through sperm banks, anonymous donations, or the husband, is commonplace. Surrogate motherhood is becoming a phenomenon that, if not common, is at least relatively well advertised. And the newer technologies of embryo transfer and embryo freezing are proving successful with respect to the goal of achieving a pregnancy brought to term.

B. Situating the Problem

When reflecting morally on the variety of technologies surrounding the process of conception and birth, it is very difficult to situate accurately the problem. A variety of perspectives have presented themselves and many moral issues thematically manifest themselves. I will present an overview of several of the concerns being addressed and then will turn to specific problems.

1. General Issues

a. Paul Ramsey

Paul Ramsey[1] has been a major, perhaps the, leading opponent of many of the technologies surrounding birth. Ramsey locates the moral discussion of most of the birth technologies under the heading of the ethics of research on human beings. His moral analysis is based on the principle of not doing harm to unconsenting research subjects.

Consequently, Ramsey makes two principal objections to birth technologies. First, they are research done on a subject who cannot consent. Second, there is the possibility of a risk of harm to the subject. Ramsey also objects to these technologies because of their possible consequences. The first consequence is the possible stigmatization of the child so conceived and all of the problems that such publicity may bring in its wake. Second, Ramsey sees these technologies as opening the door to all kinds of modifications of the human embryo, thus causing a major shift in our perception of how human beings come into existence: from procreating to manufacturing.

Ramsey's arguments against birth technologies, while based on both principles and consequences, are typically phrased in a very strident or polemic fashion. Ramsey sees nothing but problems—moral, cultural, religious, and social—coming from these technologies and his fears lead him to an overstatement of some of the issues. While not succumbing to a naive ''nature knows best'' argumentation, Ramsey's argumentation is motivated as much by his fears and mistrust of the uses of the technology as by his principled objections to it.

b. Daniel Callahan

Another perspective is provided by Daniel Callahan, who has undergone a shift in his own thinking.[2] As he notes, during the early 70's, in the initial stages of the debate, he was strongly opposed to in vitro fertilization, much on the same grounds as Ramsey. He argued that the technology was an unwarranted hazard to the fetus with no overriding moral to justify those hazards.

In his most recent analysis, Callahan argues that there are three realities that set out a general context in which we can analyze birth technologies. First, we can now do in reproductive areas what was simply thought of as impossible by earlier generations. Second, this has caused us to reconsider, in a drastic way, our relationship to nature in general, and our human nature in particular. Third, it is no longer clear what the wisdom of nature might be, much less whether there ever was such a thing.[3] Consequently, all of our moral thinking has been turned on its head and our cultural values are in a process of deep negotiation. The outcome of these discussions is presently unclear.

What is now critical for Callahan is the lack of any solid knowledge of the short or long term consequences of developments in birth technologies. Typically, what has been done is to take the risk and then to assess the consequences after the fact. Callahan concludes: "I am myself inclined to think that we ought, prudently, to let things go forward, if only because it is difficult to develop moral principles that seem powerful enough to stop matters prior to any experience whatsoever."[4]

Callahan does not intend this to mean there should be no moral analysis or that there are no relevant moral principles. Rather, because we do not know what the consequences are, on any level whatsoever, it is not clear how moral principles apply or what they tell us about how we should make decisions. Callahan thus finds himself in a position of arguing that since things have turned out reasonably well so far, there seems to be no good reason for not allowing things to go forward. He relates this, in some ways, to traditional medical practice which uses available technologies to enable people to achieve a state of ordinary capacities and possibilities.[5]

But, on the other hand, Callahan seems to be in the position of saying that since we really can't stop these situations from occurring anyway, and since there appears to be no persuasive moral argument, we might as well let these developments go ahead and see what happens. The problem with this position, one which Callahan recognizes, is that once having started on a particular course, especially in medicine or technology, it is difficult, if not impossible, to stop. Thus Callahan hints at many problems but provides little helpful moral analysis to help bring these problems to some kind of resolution.

c. Leon Kass, M.D.

A third commentator, one who interestingly enough has also changed his mind to some degree, is Leon Kass, M.D. In a footnote to a major article, Kass notes that someone observed he is becoming less passionate and more accommodating on the issue of in vitro fertilization than he previously had been.[6] Kass recognizes this shift in style, if not substance, and is unclear whether it is because of better understanding based on more reflection and experience or simply a case of familiarity breeding contempt. In any event, Kass remains bothered by many of the technologies and provides a framework in which to think about them, although he recognizes that his process of argumentation does not lead to a firm, clear conclusion.

Kass argues that the moral analysis of birth technologies, and in vitro fertilization in particular, has to begin with a moral consideration of the blastocyte. He asserts that the blastocyte, and eventually the human being, is not just a thing or stuff; it is at least potential humanity and

> as such it elicits, or ought to elicit, our feelings of awe and respect. In the blastocyte, even in the zygote, we face a mys-

terious and awesome power, a power governed by an eminent plan that may produce an indisputably and fully human being. It deserves our respect not because it has rights or claims or sentience (which it does not have at this stage) but because of what it is, now and perspectively.[7]

Because the blastocyst is predictably human and because it is analogous to an early embryo in the uterus, Kass concludes that the most sensible policy is to treat the early embryo as a pre-viable fetus and to put constraints upon its treatment similar to those on the fetus.

Kass is also concerned about the issues of motivation and the consequences of the use of these technologies. For instance, what do we mean when we say we wish to have a child of our own? The critical issue is the word "have." Does it refer to gestating and bearing or does it mean to have as a possession? Such an issue is very critical, because while it is true that there is a very deep human emotion and need for wanting and bearing children, this technology also makes it possible for the child to become yet another commodity that is available through the traditional means of the marketplace. This would represent a major shift in how we think of children and could present significant problems.

In speaking of surrogate parenting, Kass argues:

It is to deny the meaning and worth of one's body, to treat it as a mere incubator, divested of its human meaning. It is also to deny the meaning of the bond among sexuality, love, and procreation. The buying and selling of human flesh and its dehumanizing ought not to be encouraged.[8]

Another issue that Kass raises is whether the goal of providing an infertile couple with a child is a medical goal. He argues as follows:

Just as abortion for genetic defect is a peculiar innovation in medicine (or in preventive medicine), in which a disease is

> treated by eliminating the patient (or, if you prefer, a disease
> is prevented by "preventing" the patient) so laboratory fer-
> tilization is a peculiar treatment for oviduct obstruction, in
> that it requires the creation of a new life to "heal" an existing
> one.[9]

Neither this argument nor the observation that the goals of medicine are beginning to shift dramatically is unique to Kass. For many years now medicine has been providing indirect solutions for problems that are not strictly medical. While in and of itself this is not inherently problematic, nonetheless one has to recognize that providing an infertile couple with a child by means of an alternative technology does not solve the infertility problem. It also represents a shift in the purpose in medicine from one of healing to one of fulfilling the desires of individuals. Again, Kass recognizes that his argument is not foolproof nor is it totally conclusive.

2. Surrogate Mothers

One of the new phenomena arising out of the use of birth technologies is the surrogate mother. Such a surrogate can become pregnant in one of three ways: artificial insemination, in vitro fertilization, or embryo transfer. She then bears the fetus until birth at which time she returns it to the social parent(s).

This practice has given rise to many fears and concerns. It has also caused great joy for those unable to have a child in any other way or for those for whom pregnancy was dangerous. In this section we will review several problems surrounding this new development in child bearing.

a. Screening

The first issue is screening. The developing practice requires the surrogate mother to undergo a wide variety of physical and psychological tests to determine her appropriateness. The screening includes a discussion of her motivation as well as checking for the possibility of genetic or veneral disease. Screening also provides the opportunity to discuss issues that may arise during the pregnancy. One can also discuss areas of possible disagreement to avoid complications in the agreement or relation between the surrogate and the biological father.

The issue of screening raises problems similar to those discussed in the issue of artificial insemination. However, with surrogate parenting, the screening seems to be more thorough or focused. Issues of motivation as well as questions of intelligence, and general health, both for the surrogate and for her family, become exceptionally critical. This may be a time when the surrogate will learn things about herself that she did not know and perhaps would prefer not to know. There is also the issue of privacy, although one can well argue that since the surrogate has decided to perform the service, she has in effect surrendered many of her rights to privacy. However, one can still ask how thorough a surrender of privacy the surrogate makes, at least without giving express permission for some kinds of information gathering procedures, such as credit references, academic status, and criminal record.

b. Regulation of the Pregnancy

Another issue is the regulation of the pregnancy. Contracts may differ but to give a sample of the kind of issues they may contain, I will summarize some of the requirements in the contract recommended by Noel Keane, a Detroit attorney who is one of the leading legal specialists in the area of surrogate motherhood.[10] This typical contract requires the surrogate to try to become pregnant through artificial insemination and to refrain from intercourse during the time surrounding the insemination. Also, during the pregnancy, she promises to abstain from alcohol, tobacco, and dangerous drugs, and to follow the advice of her physician. The contract provides that the surrogate will not abort unless her health or that of the child is at stake and that she will abort if amniocentesis reveals a congenital defect in the child. Also the surrogate promises not to form a parent-child relationship with the child and waives the right to learn the identity of the natural father. The contract further specifies that the surrogate is to be compensated for her services and the father is to assume liability for any medical complications to the surrogate as well as contingency fees in the event of a miscarriage.

Many of these issues have been presented in the film "Paternity." While presented in a humorous fashion, the film reveals the complexity of the relationship and the degree of intrusiveness possible. "Paternity" also shows another problem: conception, pregnancy, and parenthood as entertainment.

c. Manipulation

Another issue is the potential manipulation of both the surrogate and the adoptive parents. On the one hand, parents who cannot have a child they desperately want typically undergo all kinds of diagnostic procedures and methods to attempt to establish a pregnancy. After several years of this plus an increased desire for a child, these people are ripe for a variety of manipulative or exploitative relationships. The same may be true of a surrogate mother, although data on this are unclear. The fee that is suggested, at least by Noel Keane's office, is $10,000 which is to be paid to the surrogate for her services. While $10,000 may not be much in absolute terms, it can be a powerful inducement to an individual for whom pregnancy does not appear to be onerous and who could use the money.

But the issue transcends the simple fact of pregnancy. For as we have already noted there are any number of restrictions placed upon the surrogate mother after conception has occurred. The cost of such restrictions must be factored in when evaluating whether or not these are worth the constraints on one's life style, especially upon the surrendering of the child after birth, for which I suspect no amount of counseling can prepare the individual.

d. Fetal Status

One of the interesting byproducts of the various forms of fetal therapy and the development of birth technologies has been a further clouding of the status of the fetus. On the one hand, if one approaches the fetus from the perspective of abortion legislation, one can argue the fetus has no status whatsoever. Until the time of viability, it is not considered a person. On the other hand, the clear purpose of the birth technologies is to produce a child and to enhance its conception and development. With the establishment of a contract between biological father and the surrogate mother there is the assumption that there are some things that cannot be done to the fetus. The moral basis of this appears to transcend mere rights of ownership on the part of the biological father.

New techniques in fetal surgery and treatment are further confounding the status of the fetus. If the fetus is a patient that can be treated, as is becoming more and more the case, can a mother be forced to undergo surgery for the benefit of the fetus? If the fetus is a patient and has a moral right to be treated, how does this relate to the right of the mother to abort if she so chooses? If, for example, a fetus has a birth anomaly that can be treated surgically, can the woman be restrained from aborting the child so that it can be treated? This situation may be heightened in the case of a surrogate mother who could be potentially forced to undergo surgery so that the fetus could be treated.

What is becoming clear through the development of birth technologies is that the fetus continues to be in a very ambivalent position. On the one hand, the fetus has no rights because it is not a person. On the other hand, because of the possibilities of fetal treatment and birth technologies, the fetus is becoming a patient. Because the fetus is a patient, the fetus is acquiring a right to treatment or care. In our society such rights typically are claimed only of persons and the fetus would have to be given this strong personal right's claim if the mother were to be required to undergo surgery on behalf of the fetus. Thus while pressing for the right of privacy for women has argued for a devaluation of the rights of the fetus and while the legal tradition has denied it personhood, the development of interventions on behalf of the fetus has conferred upon it the status of patient. Does the social role of patient require the moral status of a person?[11]

e. Economic Issues

A fifth issue relates to the economics of the birth technologies. Who will pay? The answers range from the possibility of federal sponsorship of research into different technologies such as in vitro fertilization to the possibility of insurance programs paying for artificial insemination, in vitro fertilization, or embryo transfers.

One also has to raise the question of what priority birth technologies have in the overall scheme in national research programs. While having a child is clearly important to those who cannot do so without some kind of medical intervention, one has to ask about the overall benefits of such research and developments of technologies in relation to the entire scheme of medical priorities for our country.

At the present time, a decision has been made that the federal government shall not fund research relating to birth technologies. Private clinics are doing in vitro fertilization and the payment for these is coming directly from the individuals. The same is typically true of surrogate parenting. Whether or not such services are to be paid for by third parties is a question that will be debated as these technologies become more successful and prevalent.

f. Risk

A sixth issue concerns the element of risk assumed by the fetus, but also by the surrogate mother. The issue of biological risks to the egg and the sperm in the process of in vitro fertilization has been proven to be minimal. Thus far, no problems have been reported with full term pregnancies initiated through in vitro fertilization. Reports of full term pregnancies initiated with frozen embryos have indicated no problems. Thus, in terms of the practice to date, there appear to be no major risks to the fetus through the use of these birth technologies. This excludes the possibility of abortion if genetic screening reveals the fetus to have a particular genetic anomaly.

The social risks to the fetus of having the potential status of illegitimacy also seem to have been resolved socially and legally. What remains to be determined is the potential effect of publicity upon the first children to be born through the various birth technologies. And while Louise Brown, the first baby to be born of in vitro fertilization, received a tremendous amount of publicity during the first several months of her life in and out of the uterus, stories on her have certainly decreased almost to the point of zero in the popular and technical press today. Thus the impact of publicity may be a function of how protective the parents are or how exploitive they may be in wanting to generate outside income by exhibiting the child. The history of the Dionne Quintuplets may alert us to problems to be avoided—especially in a media conscious society such as ours.

g. Benefits

A seventh area concerns the benefits derived from the birth technologies. Most obviously, these technologies allow people previously unable to have children to be able to have a child with some degree of genetic relatedness to them. We obviously can learn more of the reproductive process through research associated with the birth technologies. Through these technologies, physicians and scientists are also able to alleviate the experience of infertility by using technology to circumvent the impaired reproductive structure. Thus while these technologies do not cure the person of infertility they solve the problem of infertility.

h. Problems for the Future

Finally, there are several other problems that are yet to be resolved concerning the use of birth technologies. There are simply no answers available to these questions but they are important ones and deserve our most critical thinking.

a. Gender Selection.   First, many of the birth technologies give us a tremendous amount of control over gender selection. Most surveys done have indicated that individuals prefer to have the first born to be a male. It is unclear what such a disproportionate number of males may mean in our society. Such technologies would allow us to make an incredible intervention into an area in which we have absolutely no idea what that might mean.

Second, the use of birth technology to achieve gender selection is inherently discriminatory in that, as Tabithia Powledge argued, it makes survival dependent upon which gender one is. Thus the most critical reality of human beings, their existence, is dependent upon the desirability of their gender. Also the preference for males as first born may institutionalize women as second class citizens because of the traditional advantages of the first born, and especially a first born male, over all the other siblings in a family. Thus while the technologies allow a family to have the child of their choice, it is utterly unclear what new gender ratios might mean for our culture. We have no knowledge of what deleterious effects such discriminatory patterns may have on future generations.

b. Commodification of Mother and Child.   Another issue refers to the commodification of both the child and the mother. This issue has to be placed in the context of, on the one hand, the desperation and frustration of people who wish to bring a child into their family but are unable and, on the other hand, the willingness of some individuals to bear a child for couples unable to have their own. The problem that I see in this situation is that both the child and the mother can be seen as simply another object to be acquired if one has the money and access to the service. It is one thing to say that this technology can be placed at the service of a particular couple who want a child and are unable to have a child in any other fashion. I think it is quite another thing to present this as a routine means for obtaining a child. What may be at stake here is a shift that is occurring from procreating to reproducing to manufacturing.

C. Summary

Clearly at the present time there are no data to indicate that families who have had children in this fashion are abnormal or atypical or that children conceived and brought to term in this fashion have been harmed. That is precisely the reason why it is so difficult to articulate the ethical issues surrounding the whole process of the externalization of conception. One is talking about the possibility of future risks, one is talking about possible scenarios. Nonetheless, even though these questions are not clearly focused and the probabilities of future events occurring are unknown, I think it is important to raise the question of the implication of seeing conception as being part of a manufacturing process.

It is clear that for the last several decades, beginning with the advent of artificial contraception, reproduction has been separated from sexuality. That experience has proved to be exceptionally significant for several generations. Sexual roles and sexual behavior have been dramatically altered and the impact of contraception on family size is being experienced now in a very dramatic way, for example, in the absence of a college population. What will be the implications of a complete break between sexuality, the traditional means of reproduction, and the traditional family structure? We know, for example, that a couple need not be the biological parents to be good parents. The history of the traditional form of adoption has shown this to us in a very clear way. Many argue that the new birth technologies, which are described as pre-adoption technologies, will simply replicate this experience. Perhaps they will, but we may be in a radically different situation when an embryo can be selected from many different types and kinds and implanted in a woman who will carry this embryo to term for the couple who will then welcome the child into their new family after birth.[12]

10
Applications:
The Clinic and the Farm

A. Gene Therapy

One of the obvious and clear implications of many of the developments in genetics has been the hope and expectation that one would be able to intervene directly into the human gene structure either to prevent a disease or to correct an anomalous gene so that normal development could occur. In this way heritable genetic diseases could be eliminated, cured, or at least mitigated. In gene therapy, a properly functioning gene is introduced into the human organism to correct the defective one. The possibility of gene therapy presents much hope to individuals because at the present time the only forms of treatment for several serious genetic diseases are either prevention of the disease through the screening and counseling of couples before they marry, the avoidance of conception altogether, or the selective abortion of afflicted fetuses. To solve these problems, especially the last mentioned, would be a great boon for everyone.

Although recommendations for final guidelines by the National Institutes of Health are being discussed, several technical problems have been identified. First, thus far in animal work there have been generally low and variable efficiencies of gene transfer.

That is, when genes are inserted into an animal to do a specific job often the results are less than anticipated. Second, the ability of a gene to express itself depends on where it is inserted. This is a difficult process and needs further work before higher success rates can be obtained. Finally, there is the possibility that an inserted gene will alter the function of neighboring genes. This is especially problematic because while the inserted gene may perform its function as desired, it may also, through processes yet unknown, cause problems for the other genes and inhibit their performing their function.

In a recent article, Clifford Grobstein and Michael Flower have compiled precautionary guidelines based on their review of the literature.[1] These are as follows:

1. Only a disease that drastically reduces the quality or duration of life should be a candidate for somatic gene therapy.
2. A clinical trial should be conducted only if there is no alternative established therapy that is likely to yield as good or better results.
3. Investigators should be able to identify the nature of the selected genetic defect as well as the course of events leading to symptoms.
4. There should be evidence that the planned procedure for modifying the specific genetic defect is regularly safe and efficacious in comparable animal studies. This should include a demonstration that the new gene has been inserted in proper target cells; that it remains there; that it is expressed appropriately (in other words, produces proper quantities of its product); and that it does no harm to target cells or, inadvertently, to nontarget cells.
5. All established procedures for the ethical conduct of human clinical trials should be followed.
6. The protocol should be so planned that even if therapy is not achieved its subsequent success will be more likely, i.e., "shots in the dark" should not be attempted.

These guidelines help focus attention on both the clinical and ethical problems involved in gene therapy. While they restrict some forms of research and make a strong argument for a cautionary approach toward implementation of gene therapy, nonetheless they present a context in which gene therapy can be implemented from both technical and ethical perspectives.

B. Agricultural Application

Most of the basic research in genetics has occurred in both plants and animals and, to conclude this survey of ethical issues, I wish to include some comments on agricultural developments.

A recent article in *The New York Times* indicated a shift on the federal level from conventional plant and animal breeding to the use of genetic engineering and other biotechnologies for plant and animal breeding programs.[2] This includes a quadrupling of the research budget so that research using cloning and gene splicing techniques, the transfer of embryos, the regeneration of plants from tissue cultures, and the synthesis of various hormones can be examined. These new technologies are going to be expanded to make new hormones to produce meatier animals for human consumption, to develop entities that can be placed in the soil to break down harmful chemicals before they can contaminate the ground water supply, to enable a greater variety of fruits and vegetables to utilize the sun's energy to become a more appropriate food source, and to continue research on a long desired issue: enabling plants to draw nitrogen from the air. This will accomplish two main goals: first, it will reduce the amount of fertilizers needed which will decrease agricultural costs to some extent; second, it will free up oil supplies because these are presently used to produce fertilizers.

Long range research plans include: the enhancement of fruits, vegetables, and animals for human consumption; the reduction of deleterious impacts upon the environment; a more efficient use of the resources that we have at our disposal. Such research carries with it great promise and expectations for relieving many of the food shortages as well as providing new forms of food for human consumption.

Another major area in the use of biotechnologies has to do with the creation of genuine chimeras. The classic chimera was a mythological animal composed of three different parts: a lion on the front, a female goat in the middle, and a serpent on the rear. Mythology and science fiction allowed us to become familiar with and even captivated by different animal forms. We also have E.T. and all the Star Wars creatures who have delighted us. Yet, most of us would be unprepared for the report in a recent edition of *Nature*[3] that goat-sheep chimeras could be biologically produced. This is an animal containing the tissues of two distinct genetic types. Such work had been done earlier by combining different species of mice and even the rat. It is only more recently that diverse embryos from mammals were used.

In a recent article describing some of these processes, Bernard Dixon raises several reasons why such research was initiated and what some of its implications are.[4] The two primary motivations for this research are: understanding the processes by which diverse cells arise and interact during traditional development; second, to determine whether success in removing the reproductive barrier between species would facilitate both the transfer of embryos between species and the production of previously unachievable hybrids. This might allow rare animals to be bred by using commoner animals as surrogate mothers. Thus there is a theoretical gain to be made in terms of further knowledge of the process of embryonic development and there is a practical goal of saving many animals from extinction if they can be bred in other animals. Conservationists in particular find this extremely exciting since there are many species that are being kept alive solely in zoos. The removal of such natural barriers to reproduction could help conservationists breed rare species.

Even given these benefits, a rather critical question remains: what about the creation of chimeras involving human beings? The traditional answer is that it is unrealistic to anticipate such research or that it is too early even to think of it. Dixon comments on this issue in very ominous tones.

> Moreover, given that one motive for bringing sheep and goats together was to explore the possibility of genuine hybridization, why should we suppose that the reproductive barriers between humans and "lower" primates are insuperable? Despite the fact that experimentation along these lines seems unthinkable or distasteful there is no basis whatever for believing that it will never happen.[5]

What Dixon is doing here is showing how two elements—genuine research interest and the technological imperative—become merged and issues or projects that are unthinkable or problematic will be done simply because of the pressures of the research itself.

C. Summary

While it is true that many desirable and even necessary benefits can be derived for human beings from increases in the study of both animal and plant genetics, it is also true that research in these areas will continue to allow us to recombine species in new ways. While such efforts may be in the future, now is the time to begin thinking about whether we want to create chimeras or hybrids with both animal and human characteristics.

One of the major fears that I have always had in this area is the frequency of the suggestion that the appropriate use of such hybrids will be to do the drudgery work that human beings typically do not like. The creation of a slave population appears frequently in science fiction and some ethical literature, more than does the creation of human beings with characteristics such as kindness, compassion, friendliness, and a desire for community. My fear is not necessarily that we will do these things, but that the creation of such new kinds of slaves are among the first things we typically think of in this area. While we should look to future goods, we must also remember that a noble or good purpose does not justify any means. The time to consider the ethical and social considerations of the technology is now, before their implementation.

11
Shaping the Debate:
Positions on Genetic Engineering

A. Overview of the Issues

One can hardly pick up the paper on any random day without encountering yet some new development that has emerged from genetic engineering. On April 11, 1984, *The New York Times* reported the first baby born of a frozen embryo. After the ovum was fertilized in the laboratory with the husband's sperm, it was then frozen for two months before being implanted into a woman's prepared uterus. And, on June 5, 1984 another story in *The New York Times* reported that bits of genetic material from the quagga, an extinct relative of the zebra, had been found and reproduced in the laboratory. The technology of cloning allowed the genetic material to be reproduced so it could be studied. While the possibility of recovering enough genetic information to bring an extinct species back to life is exceptionally remote, nonetheless the technology of cloning will allow scientists to study extinct species to learn more about their relationship with other species living today.

Stories like these continually appear in the popular media as well as in scientific and professional journals. Developments in genetics are coming at a fast and furious rate. The biotechnology industry continues to grow and new relationships between scientists, universities and industry appear to be a fixed part of the bioengineering landscape. The agricultural research service of the federal government is planning on quadrupling the budget spent on research employing various technologies of genetic engineering. These technologies should allow breakthroughs in both agriculture and animal husbandry.

We can expect even more developments in genetics in the future. This raises the necessity of evaluating the biological, social, and cultural aspects of genetic engineering. And it is precisely in this area that the most difficult problems arise. As has been indicated throughout this book, many of the problems are difficult to analyze because we are dealing with future possibilities and with attempts to evaluate different models of the future. Also it is clear that many of the developments in genetic engineering, especially those related to birth technologies, bring tremendous short term benefits. What is unclear is what the long term benefits and risks, if any, may be and how they will affect our society. The fact that the questions are difficult to analyze does not excuse us from ignoring them. We must continue to analyze the questions as responsibly as we can and to be concerned about the implications of our tremendous powers. Otherwise we may live in a future of our wants but not our needs.

B. Focusing on the Issues

A number of questions have been raised about genetic engineering. Some of these questions are complementary; others move clearly in opposing directions. In this section I want to present an overview of questions and perspectives that have been raised about various aspects of genetic engineering to show the different kinds of questions being discussed and to present the various agenda being brought to the issues of genetic engineering.

1. An Alternative Agenda

When the first major debates about recombinant DNA began in the mid '70s, Jeremy Rifkin led a coalition that raised several critical questions about recombinant DNA.[1] While Rifkin and his followers did not answer these questions, nonetheless the questions are significant and I wish to list them as a means of aiding an evaluation of some of the issues involved in genetic engineering from an ethical, scientific, and political perspective. The questions are as follows.

1. What are the moral, ethical and theological implications involved in the artificial creation of new forms of life?
2. Does humankind have the necessary knowledge and wisdom to successfully circumvent the evolutionary process without creating long term and catastrophic eco-disasters?
3. Will decisions on the creation of new life forms be left to the discretion of individual scientists, bureaucrats, and corporations?
4. Do corporations have the right to create and patent new forms of life for profit?
5. Is there any justification whatsoever for any secret recombinant DNA research to continue in commercial laboratories?
6. What are the moral and political implications involved in the use of genetic improvement techniques on humans?
7. How can this new area of experimentation be carefully monitored if it is allowed to continue in any laboratory in the country?
8. Who has the right to decide if the risks of accidental contaminations and epidemics are insignificant?
9. How will the entire American public be involved in decisions we must now make as a society?
10. Due to the long term consequences and risks of this work, is there any reason at all why there should not be an immediate moratorium on recombinant DNA experimentation until our society has engaged in a full national debate on this issue?

These questions were raised in 1976 at the beginning of the debate about recombinant DNA research. In the eight years since these questions were raised, several issues have been resolved. The courts have permitted various life forms to be patented and licensed for profit. This led to many bio-technical corporations being founded to do research and applications in genetics and other areas using the recombinant technologies. We also have a fair amount of experience now with the safety of the technology. The guidelines in place have established a level of safety and no accidents with recombinant DNA have been reported. What has not been resolved, and continues to be a problem, is the application of the technology to create new life forms and how, if at all, that will be controlled.

2. *National Conference of Catholic Bishops*

In May of 1977 the National Conference of Catholic Bishops issued a statement on recombinant DNA research.[2] The bishops saw this statement primarily as an educational one but also identified the recombinant DNA issue as a paradigm for discussing other scientific issues.

The bishops raised several questions that related to the potential risks involved in the research—substantive or marginal—and who should judge the acceptability of those risks. The bishops also encouraged public debate about the issue. Finally, there was concern with the developing guidelines with respect to their enforceability, their appropriateness, and whether such legislation was in fact the best way to monitor this type of research. The final question that the bishops raised was the extent to which our country ought to advance in genetic engineering.

Another theme in the bishops' statement is the issue of freedom of scientific inquiry and the role of the public in science policy. Scientists argued that any attempts to stop or regulate recombinant DNA research was a primary infringement upon their freedom of thought and research. The bishops argued that the knowledge gained in recombinant DNA research ought not to be at the expense of other important values in the human community. They urged, therefore, a note of caution and prudence in beginning to implement a major research project.

The bishops' moral analysis recognizes that while we are obligated to avoid harm, nonetheless, we have no necessary mandate to accomplish all of the possible goods that we can. Therefore we need to evaluate whether future generations will benefit by our pursuit of interesting but possibly unnecessary areas of scientific research. They also argue that we need to avoid being trapped by the technological imperative which suggests that if we can do something we ought to do it. This is coupled with caution against using a strictly utilitarian mode of thinking since this can obscure other significant values.

Finally, the bishops argue that recombinant DNA research should not be viewed solely from the perspective of a risk-benefit calculus. This methodology has the potential to obscure other values at stake. Nonetheless, the bishops argue that the urgency of the problems ought not to limit reflection on the purpose and implications of DNA modification, the effect of this type of genetic research on our understanding of ourselves and of our relation to nature, and the correlation between the scientific advance possible through DNA research and human progress as judged by a variety of criteria.

The bishops basically caution Americans not to rush wholeheartedly into new ventures without thinking through all of the consequences. They also suggest that there are values other than scientific progress and that part of our ethical responsibility as citizens is to think through the major values of our society. We then need to put the benefits to be derived from science and technology into that framework rather than isolating scientific and technological values from their social context.

The bishops see the issue of recombinant DNA research primarily in terms of a conflict of values and urge that attempts be made to resolve the question with respect to priorities among values rather than exclusively on utilitarian or risk-benefit grounds. Through this method the bishops hope we can escape the trap of being guided by the technological imperative.

3. In Vitro Fertilization

Guidelines set in place by the Department of Health and Human Services prohibit the support of research involving in vitro fertilization until reviewed by the Ethics Advisory Board. In 1977, HHS received such an application and, in May 1978, the Ethics Advisory Board reviewed the research proposal and released a report on their findings a year later. First, I will present their conclusions.[3]

1. The Department should consider support of carefully designed research involving in vitro fertilization and embryo transfer in animals, including non-human primates, in order to obtain a better understanding of the process of fertilization, implantation and embryo development, to assess the risks to both mother and offspring associated with such procedures, and to improve the efficacy of the procedure.

2. The Ethics Advisory Board finds that it is acceptable from an ethical standpoint to undertake research involving human in vitro fertilization and embryo transfer provided that:

A. If the research involved human in vitro fertilization without embryo transfer, the following conditions are satisfied:

(1) The research complies with all appropriate provisions of the regulations governing research with human subjects (45 CFR, 46).

(2) The research is designed primarily: (a) to establish the safety and efficacy of embryo transfer and (b) to obtain important scientific information toward that end not reasonably attainable by other means.

(3) Human gamites used in such research will be obtained exclusively from persons who have been informed of the nature and the purpose of the research in which sets of materials will be used and have specifically consented to such use.

(4) No embryo will be sustained in vitro beyond the stage normally associated with the completion of implantation (fourteen days after fertilization).

(5) All interested parties and the general public will be advised if evidence begins to show that the procedure entails risks of abnormal offspring higher than those associated with natural human reproduction.

B. In addition, if the research involves embryo transfer following human in vitro fertilization, embryo transfer will be attempted only with gametes obtained from lawfully married couples.

3. The Board finds it acceptable from an ethical standpoint for the Department to support or conduct research involving human in vitro fertilization and embryo transfer, provided that the applicable conditions set forth in conclusion 2 are met. However, the Board has decided not to address the question of the level of funding, if any, which such research might be given.

4. The National Institute of Child Health and Human Development and other appropriate agencies should work with professional societies, foreign governments and international organizations to collect, analyze and disseminate information derived from research (in both animals and humans) and clinical experience throughout the world involving in vitro fertilization and embryo transfer.

5. The Secretary should encourage the development of a uniform model law to clarify the legal status of children born as a

result of in vitro fertilization and embryo transfer. To the extent that funds may be necessary to develop such legislation, the Department should consider providing appropriate support.

The approach taken by the Ethics Advisory Board is one of gradualism which stresses that research must continue on the animal level so that individuals can obtain a better sense of what risks and benefits, if any, are associated with the procedure. The Board chose not to recommend funding such research and the Department of Health and Human Services accepted that recommendation. Thus all research in the area being carried out, as well as the development of various clinics in the country doing in vitro fertilization or embryo transfer, is funded from the private sector.

At issue in the ethical considerations of the Ethics Advisory Board were problems related to the moral status of the embryo, the priority of funding such research when other health needs are unmet, issues concerning the safety of the research for both the fetus and the mother, and the possibility of linking this method of conception with other forms of genetic intervention. These ethical considerations caused the Board to be conservative in its recommendation on funding, but also allowed certain forms of research to go forward which were designed to test the safety and efficacy of the procedure itself.

Thus after these basic questions are resolved, it will be appropriate to raise the issue of the possibility of the funding of research connected with embryo transfer. The problem with this is the requirement that such research be reviewed by the National Ethics Advisory Board. There is no such board in existence at present and it is unclear whether one would be established simply to review one or the other research proposals. On the other hand, as I noted above, much of the work on in vitro fertilization and embryo transfer is occurring in the private sector and the results thus far have not shown any major risks to either mother or fetus.

4. Splicing Life: A Report of the President's Commission for the Study of Ethical Problems in Medicine and Biomedical and Behavioral Research

The specific report was developed by the President's Commission in response to a letter written in 1980 by leaders of Protestant, Catholic and Jewish organizations. These religious leaders focused on ethical issues relating to forms of genetic engineering, the patterning of new life forms, and the control of these life forms, especially through the commercialization of genetically engineered materials. The Commission spent two years reviewing many of the issues involved in genetic engineering and issued in November of 1982 the report *Splicing Life*. This Commission made several findings which I will summarize.[4]

First, the Commission noted that much public concern about genetic engineering reflected an anxiety that such research might remake human beings. The Commission finds that these claims appear to be exaggerated. While recognizing that some of the technologies may challenge deeply held feelings about human life as well as provide new methods of curing illness, nonetheless the new knowledge is a celebration of human creativity and a reminder to act responsibly.

Second, the Commission notes that these engineering techniques are advancing extremely rapidly. For example, artificially inserted genes have functioned in succeeding generations of mammals.

Third, these techniques are already demonstrating their great potential value for human well-being. The decrease in suffering is the ethical reason for continued support. Many of these techniques will be reviewed by Institutional Review Boards which will allow a review of the risk-benefit ratio. This mechanism, already in place, will allow review of procedures and an evaluation of the safety issues.

Fourth, the Commission noted that while many uses of genetic engineering resemble traditional forms of diagnosis and treatment and therefore should not cause any major problem, close scrutiny may be appropriate for interventions that would bring about a heritable genetic change. Also, treatments which would enhance qualities considered as normal may also be subject to further consideration. This is because once enhancement of a particular quality is attempted, we are moving very clearly in the direction of trying to achieve a perfect human being. This can raise serious problems.

Fifth, the Commission notes that many people object to genetic technology because this is perceived as "playing God." While the Commission argues that the scientific procedures in question do not inherently appear to be inappropriate for human use, nonetheless the strong potential contained within these technologies ought to remind us that we need to evaluate and monitor continuously what we do with genetic technology.

Sixth, the Commission notes that there is a good record of safety with respect to gene splicing technology. This has come about as the result of initial concern, careful monitoring, development of appropriate technology, and the cooperation of the scientists involved. The working assumption is that people who would assume that this research is not safe have the burden of proving this. The working assumption today is that gene splicing technology is safe, based on the record.

Seventh, the Commission notes the work and role of the Recombinant DNA Advisory Committee, RAC, and how it has continued to develop its guidelines for different types of genetic engineering. The Commission argues that now is the time to broaden its area and to include issues raised by the application of the techniques of gene splicing rather than simply the early issue of the unintended exposure from laboratory experiments.

Eighth, the Commission recommends that such a process of scrutiny carried out under the auspices of the RAC include individuals from many different backgrounds including congressional and executive branch agencies, but also scientific and academic associations, members of industrial and commercial organizations, ethicists, attorneys, and leaders in the area of education and religion, as well as members of the general public.

Finally, the Commission argues that there is need for an oversight body based upon the "profound nature of the implications of the gene splicing as applied to human beings, not upon any immediate threat of harm." Thus, the Commission recognizes that the major issue is the enormity of the implication of the technology when applied to human beings. The primary issue of safety appears to have been reasonably resolved; the other major problems of the implications of this technology for human well-being are yet to be determined. For this reason the Commission appropriately recommends continuous monitoring of the research.

The President's Commission has shown a fair amount of prudence, in my opinion, in terms of these recommendations and evaluations. The Commission recognizes that there is a history of safety and responsibility in dealing with the whole issue of genetic engineering and gene splicing. This is due in large part to the cooperation among scientists and between scientists and the federal government. When safety issues were first discussed, scientists recognized that they had a primary role in developing leadership to evaluate the safety issues and to set the whole tone of the debate by their own example. The Commission recognizes those efforts and, because there has been a good safety record, acknowledges that there can be some relaxation of some regulations on genetic engineering.

On the other hand, the Commission is aware of the tremendous potential that genetic engineering has in terms of human welfare. Thus it prudently recommends monitoring of new areas of research as well as a continuing evaluation of the application of genetic engineering to various areas. The Commission refused to be swayed by scare tactics but also recognized the reality of the concerns being expressed by many individuals. Thus the Commission set forth the framework that can help evaluate and monitor continuing developments in genetic engineering.

5. Foundation on Economic Trends: Concerns for Theological Issues in Genetic Engineering

The Foundation on Economic Trends, an organization founded by Jeremy Rifkin, issued in June of 1983 a resolution on theological issues in genetic engineering.[5] This resolution was both signed and sponsored by sixty-three religious leaders and scientists representing every spectrum of the political and religious scene.

The basic position of the resolution is that "efforts to engineer specific genetic traits into the germ line of the human species should not be attempted." This resolution has two factors behind it. The first concerns technical issues. The resolution recognizes that molecular biologists have succeeded in altering the sex cells of a mammalian species through genetic engineering. New advances in genetic engineering raise the possibility of altering the human species. Engineering specific genetic traits into the sperm, egg, or embryo of a human being represents a basic alteration in the way a human being may be formed.

The second concern is for an evaluation of the moral issues. The redesign of the human species through genetic engineering irreversibly alters the composition of the gene pool for future generations. Genetic engineering will necessitate decisions about which genetic traits should be programed into the human gene pool and which should be eliminated. No individual, group of individuals, or institution can legitimately claim the authority to make that decision on behalf of the rest of the human species either now or in the future.

Standing behind this resolution is concern about the price to be paid for embarking on the goal of perfecting the human species. Rifkin argues that there is an ecological price to pay in that eliminating what are perceived to be bad genes could lead to a narrowing of diversity of the gene pool. Also in attempting to perfect the human being by focusing on good genes, the human genetic pool is similarly narrowed. Rifkin is also bothered by the past history of eugenics and the definition of certain groups as inferior, as well as the violation of the civil rights of individuals who are mentally ill or retarded. Rifkin is concerned, as are many others, with the price that is to be paid for attempting to perfect the human species. There are issues about the specifications for perfection and there are significant issues in terms of who are the decision makers.

While Rifkin is perceived by many to be raising questions in an alarmist fashion, nonetheless many of these issues are critical and need a reasonable response from the scientific community, as well as from various funding agencies and the federal government. It is important, for example, to understand the effect of eliminating or converting certain genes within the gene pool. What are the criteria for a perfect human being and what implications will this have? And, finally, who will be the decision makers within our society and how will they be appointed? Rifkin has done a major service by continuing to call these issues to our attention. On the other hand one can make a reasonable argument that the alarmist tone underlying the resolution will not help his cause.

C. Conclusion

The one firm conclusion that can be drawn in the area of genetic engineering is that events continue to outpace our capacity to think through the issues or to have reasoned philosophical or theological discussions on the problems. Events and technologies continue to overtake us and problems continue to mount.

In May 1984, the journal *Science* reported on several major patent disputes that are occurring between biotechnology companies over the use of almost identical methods to solve similar problems. Billions of dollars as well as control of the industry are at stake.[6]

Another major dispute has to do with an experiment that would release into the environment genetically engineered bacteria on plants to determine whether these bacteria improve the plants' resistance to frost damage. A Federal Appeals Court in Washington upheld a lower court decision to halt this experiment proposed by researchers at the University of California. On the other hand the previously discussed recombinant DNA Advisory Committee (RAC) approved a request from Advanced Genetic Sciences, Inc. of Greenwich, Conn. to spray genetically engineered bacteria on plants to test the same problem. The RAC also endorsed a request by Cetus Madison Corporation of Wisconsin to test in the field plants genetically engineered for resistance to disease.

The allowed research testing resistance to frost damage is identical with the experiment proposed by the University of California which is prohibited. The basis for this contradictory position is that biotechnology companies do not receive their support from the National Institutes of Health and, therefore, are not required by federal regulations to have their experiments reviewed and approved by the NIH panel. Typically these companies do submit their protocols for review to the NIH, but its approval is not mandatory before testing a product in the environment. Thus the issues of the control of genetic engineering and the relationship between industry and academia are raised again.[7]

Finally, in an event that once again proves that truth is stranger than fiction, several news stories in June of 1984 reported on the existence in Australia of two frozen embryos who were orphaned when the married couple from whom the egg and sperm were obtained died in a plane crash.[8] The embryos were conceived in vitro and later frozen. The debate was over whether to attempt to thaw out the embryos and implant them so that they can be gestated or to destroy them. Complicating the issue is the fact that these embryos, if brought to term, possibly stand to inherit the multi-million dollar estate of their now dead parents. The issue of inheritance may be moot because of a later discovery that the sperm donor was not the husband.

In Australia, as well as everywhere else, there are absolutely no guidelines on any level to determine what the fate of a frozen embryo might be in the event of the death, divorce, or other problems that may occur between a couple who have chosen to conceive a child in vitro and then have the embryo frozen until a later date. This case raises clearly the issue of the moral status of the embryo as well as showing the necessity of some kind of guidelines to determine what should be done to embryos in such cases. Many countries are now beginning to develop such guidelines in the light of this case, but these are in the initial phases of discussion.

We have seen over and over again in this book that genetic engineering presents us with tremendous potential with respect to agriculture, therapeutic implications, and human well-being. On the other hand, there has been a constant underlying suggestion that there are areas that we have not thought of, there are areas that we have not examined, and there are implications that may have significant impact on all of our lives. Yet there appears to be a continuing gap between the developments in genetics and concern for issues of public policy or an examination of the ethical issues involved.

While the Presidential Commission's report, *Splicing Life,* is correct in its assessment that at the present time no clear and present dangers appear to be connected with genetic engineering and that no deleterious applications of the technology are likely, nonetheless I would argue that we need to be alert to the consequences of what we are doing now catching up to us. We need to begin thinking about the impact of birth technologies on the family, what parenting means in our contemporary situation, and the possibility of environmental impact. While the tone and style of Jeremy Rifkin is alarmist and quite shrill, nonetheless many of the issues that he raises are worthy of our very serious consideration. In his new book, *Algeny,*[9] Rifkin focuses on our current drive to change the essence of a living thing by transforming it from one state to another: the engineering of the perfect organism. While one can disagree with several of the conservative conclusions that Rifkin draws from his study of bio- and genetic engineering, one nonetheless senses that the underlying theme of the book of a wariness about some of the applications of this technology is a point well taken.

The future of genetic engineering holds great promise. But we must always remember that genetic engineering is controlled by human beings who operate on the basis of values and preferences. Science and genetics are not value free entities that have lives of their own. Science and studies in genetics and biology have given us a fascinating insight into our world. They have shown us technologies that can change who we are. But these changes come about because of the desires of human beings. It is these desires and their consequences that we must examine thoroughly as we step into a new world. Only through such examination, difficult and problematic as it is, can we be responsible to ourselves, to our ancestors and to our environment.

Notes

Chapter 2

1. Pope John Paul II, *Redemptor Hominis,* # 15.
2. Daniel Callahan, "The Moral Career of Genetic Engineering." The Hastings Center *Report* (April, 1979), pp. 9–10.
3. June Goodfield, *Playing God: Genetic Engineering and the Manipulation of Life.* New York: Random House, 1977.
4. Daniel Callahan. "Ethical Responsibility in Science in the Face of Uncertain Consequences." *Annals* of the New York Academy of Science. 265: pp. 2–4.

Chapter 3

1. Key Dismukes, "Recombinant DNA: A Proposal for Regulation." The Hastings Center *Report* (April, 1977), p. 27.
2. Daniel Callahan, "Ethical Responsibility in Science in the Face of Uncertain Consequences," *Op. cit.,* p. 10.
3. David H. Smith, Scientific Knowledge and Forbidden Truths." The Hastings Center *Report* (December, 1978), pp. 30–35.

Chapter 4

1. Robert T. Francoeur. "We Can—We Must: Theological Reflections on the Technological Imperative." *Theological Studies* (September, 1972), p. 429.

2. Karl Rahner. "Christianity and the 'New Man'." *Theological Investigations*. Vol. V. Baltimore: Helicon Press. Pp. 135ff.

3. Joseph Fletcher. "Indicators of Humanhood: A Tentative Profile of Man." The Hastings Center *Report* (November, 1972), p. 4ff.

4. Joseph Fletcher. "Four Indicators of Humanhood—The Debate Matures." The Hastings Center *Report* (February, 1979) p. 4ff.

5. A. MacIntyre. "Seven Traits for the Future." The Hastings Center *Report* (February, 1979) p. 5ff.

Chapter 5

1. J. B. Schneewind. "Sociobiology, Social Policy and Nirvana." In M. S. Bregory, A. Silvers, and D. Sutch, eds., *Sociobiology and Human Nature*. San Francisco: Jossey-Bass, 1978. Pp. 234ff.

2. Arthur Caplan. "Genetic Aspects of Human Behavior: Philosophical and Ethical Issues." *The Encyclopedia of Bioethics*. New York: The Free Press, 1978. Vol. II. P. 541.

Chapter 6

1. Richard McCormick, S.J. "Notes on Moral Theology." *Theological Studies* (March, 1979) pp. 108–109.

2. Karl Rahner. "The Problem of Genetic Manipulation." *Theological Investigations*. Vol. IX. New York: Seabury Press, 1975. P. 236.

3. Leon Kass. " 'Making Babies' Revisited." *The Public Interest*. (Winter, 1979), pp. 41–54.

4. Karl Rahner. "Experiment: Man." *Theology Digest* (February, 1978), p. 67.

5. C. Boorse. "What a Theory of Mental Health Should Be." *Journal of the Theory of Social Behavior*. Vol. 6, pp. 61–84.

Chapter 7

1. Theodosius Dobshansky. "Man and Natural Selection." *American Scientist*. Vol. 49. Pp. 205–299.

Chapter 8

1. Helpful articles are: Stanley N. Cohen. "The Manipulation of Genes." *Scientific American.* Vol. 223, pp. 25–33. Stanley N. Cohen. "Recombinant DNA: Fact and Fiction." *Science.* Vol. 195. pp. 654–657. R. Roblin. "Reflections on Issues Posed by Recombinant DNA Molecule Technology." *Annals* of the New York Academy of Science. Vol. 256. Pp. 59ff.

2. Willard Gaylin. "The Frankenstein Factor." *The New England Journal of Medicine.* 12 September 1977.

3. Katherine Barton. "Academic Research and Big Business: A Delicate Balance." *The New York Times Magazine.* 11 September 1983. Pp. 62ff.

Chapter 9

1. Paul Ramsey. "Manufacturing Our Offspring: Weighing the Risks." The Hastings Center *Report.* Vol. 8. Pp. 7–9. Confer also Ramsey's earlier works "Shall We Reproduce?" *Journal* of the American Medical Association. Vol. 220. Pp. 1346–50 and 1480–85.

2. Daniel Callahan. "Ethics and Reproductive Biology." in Nora K. Brill, ed., *Who Decides? Conflicts of Health Care.* NJ: The Humana Press. Pp. 169ff.

3. *Ibid.,* p. 171.

4. *Ibid.,* p. 177.

5. *Ibid.,* p. 176.

6. Leon Kass. " 'Making Babies' Revisited." In Thomas A. Shannon, ed., *Bioethics,* NJ: Paulist Press. 1981 Revised Edition. Pp. 445ff.

7. *Ibid.,* p. 451.

8. *Ibid.,* p. 458.

9. *Ibid.,* p. 464.

10. Noel P. Keane. "The Surrogate Parenting Contract." Available through the offices of Noel Keane, 930 Mason, Dearborn, MI 48124.

11. This issue will be more confounded if recent research on intra-uterine learning proves correct. Studies on sound discrimination—using recordings of intrauterine sounds, postbirth preferences for books read to the fetus and discrimination of the maternal voice—may provide a whole new perspective at least on the viable fetus. Cf. Gina Kolata. "Studying Learning in the Womb." *Science.* Vol. 225. Pp. 302–303.

12. Articles of interest include: Noel P. Keane. "Legal Problems of Surrogate Motherhood." *Southern Illinois University Law Journal.* 1980: 147ff. Philip Parker. "Surrogate Motherhood: The Interaction of Litigation, Legislation, and Psychiatry." *International Journal of Law and Psychiatry.* Vol. 5. P. 341ff. Clifford Grobstein. "External Human Fertilization." *Scientific American.* Vol. 240. Pp. 57ff. Clifford Grobstein et al. "External Human Fertilization: An Evaluation of Policy." *Science.* Vol. 222. Pp. 127ff. Thomas A. Shannon. "The Case Against Test Tube Babies." *National Catholic Reporter.* 11 August 1978. "Surrogate Mothers: The Case For and Against." John Robinson and Herbert T. Krimmell. The Hastings Center *Report.* October. 1983. Charles F. Westoff and Ronald R. Rindfuss. "Sex Preselection in the United States: Some Implications." *Science.* Vol. 184. Pp. 633ff.

Chapter 10

1. Clifford Grobstein and Michael Flower. "Gene Therapy: Proceed with Caution." The Hastings Center *Report.* April 1984. See also the recent survey article: P.T. Rawley, "Genetic Screening: Benefit or Menace?" *Science,* Vol. 225: 138–144.

2. Bill Keller. "Key Farm Laboratory Plans 'Overdue' Shift to Genetics." *The New York Times.* 29 May 1984. P. A1. Confer also Kim McDonald. "Rapid-Growth Genes Could Yield 'Super Livestock.' " *The Chronicle of Higher Education.* 8 February 1984, p. 1.

3. *Nature,* Vol. 307: 634–38.

4. Bernard Dixon. "Engineering Chimeras for Noah's Ark." The Hastings Center *Report.* 2 April 1984. Pp. 10–12.

5. *Ibid.,* p. 12.

Chapter 11

1. "An Alternative Agenda: The Ten Key Questions. People's Business Commission. 1346 Connecticut Ave., NW. Washington, DC 20036.

2. Statement on Recombinant DNA Research. Bishops' Committee for Human Values. National Conference of Catholic Bishops. 1977. The United States Catholic Conference. 1312 Massachusetts Ave., NW. Washington, DC 20005.

3. Ethics Advisory Board. Report and Conclusion. HEW Support of Research Involving Human In Vitro Fertilization and Embryo Transplant. 4 May 1979. (Out of Print)

4. *Splicing Life.* Report of the President's Commission for the Study of Ethical Problems in Medicine and Biomedical and Behavioral Research. Superintendent of Documents. U.S. Government Printing Office. Washington, DC 20402.

5. Available from the Foundation on Economic Trends. Suite 1010, 1346 Connecticut Ave., NW. Washington, DC 20036.

6. Jeffrey L. Fox. "Gene Splicers Square off in Patent Courts." *Science.* 224: 584–86.

7. Colin Norman. "Judge Halts Gene-Splicing Experiment." *Science* 224: 962–63. Colin Norman. "Appeals Court Upholds Legal Block on Experiment." *Science.* 224: 1079. Kim McDonald. "NIH Approves Genetics Experiment Similar to One that Judge Barred." *"The Chronicle of Higher Education.* 13 June 1984, p. 7.

8. *The Evening Gazette.* 18 June 1984. Worcester, MA, p. 1. Sandra Blakeslee. "New Issues in Embryo Case Raised Over Use of Donor." *The New York Times.* 21 June 1984, p. A16. This article states that the sperm donor was not the husband in the case but an anonymous donor.

9. Jeremy Rifkin. *Algeny.* NY: Penguin Books, 1984.

Bibliography

Barton, Katherine. "Academic Research and Big Business: A Delicate Balance." *The New York Times Magazine*. 11 September 1983. Pp. 62ff.

Blakeslee, Sandra. "New Issues in Embryo Case Raised Over Use of Donor." *The New York Times*. 21 June 1984. P. A16.

Boorse, C. "What a Theory of Mental Health Should Be." *Journal of the Theory of Social Behavior*. Vol. 6: 61–84.

Callahan, Daniel. "Ethical Responsibility in Science in the Face of Uncertain Consequences." *Annals* of the New York Academy of Science. 265: 2–4.

Callahan, Daniel. Ethics and Reproductive Biology." In Nora K. Brill, ed. *Who Decides? Conflicts of Health Care*. NJ: The Humana Press.

Callahan, Daniel. "The Moral Career of Genetic Engineering." The Hastings Center *Report*. April, 1979. Pp. 9–10.

Caplan, Arthur. "Genetic Aspects of Human Behavior: Philosophical and Ethical Issues." *The Encyclopedia of Bioethics*. NY: The Free Press. 1978. Vol. II, p. 541.

Cohen, Stanley. "Recombinant DNA: Fact and Fiction." *Science*. Vol. 95: 654–57.

Cohen, Stanley. "The Manipulation of Genes." *Scientific American*. Vol. 223: 25–33.

Colin, Norman. "Judge Halts Gene-Splicing Experiment." *Science*. Vol. 224: 962–63.

Colin, Norman. "Appeals Court Upholds Legal Block on Experiment." *Science*. Vol. 224: 1079.

Dismukes, Key. "Recombinant DNA: A Proposal for Regulation." The Hastings Center *Report*. April, 1977, p. 27.

Dixon, Bernard. "Engineering Chimeras for Noah's Ark." The Hastings Center *Report*. April, 1984. Pp. 10–12.

Dobzhansky, Theodosius. "Man and Natural Selection." *American Scientist*. Vol 49: 205–209.

Fletcher, Joseph. "Four Indicators of Humanhood—The Debate Continues." The Hastings Center *Report,* February, 1979, p. 4ff.

Fletcher, Joseph. "Indicators of Humanhood. A Tentative Profile of Man." The Hastings Center *Report,* November, 1972. P. 4ff.

Fox, Jeffery. "Gene Splicers Square off in Patent Courts." *Science*. Vol. 224: 1079.

Francoeur, Robert. T. "We Can—We Must: Theological Reflections on the Technological Imperative." *Theological Studies,* September, 1972, p. 429.

Gaylin, Willard. "The Frankenstein Factor." *The New England Journal of Medicine*. 12 September 1972.

Goodfield, June. *Playing God: Genetic Engineering and the Manipulation of Life*. NY: Random House. 1979.

Grobstein, Clifford and Michael Flowers. "Gene Therapy: Proceed with Caution." The Hastings Center *Report*. April, 1984.

Grobstein, Clifford. "External Human Fertilization." *Scientific American*. Vol. 240: 57.

John Paul II. *Redemptor Hominis*.

Kass, Leon. " 'Making Babies' Revisited." *The Public Interest*. Winter, 1979.

Keane, Noel. "Legal Problems of Surrogate Motherhood." *Southern Illinois University Law Journal*. 1980: 147.

Keane, Noel. "The Surrogate Parenting Contract." Available from Mr. Keane at 930 Mason, Dearborn, MI 48124.

Keller, Bill. "Key Farm Laboratory Plans 'Overdue' Shift to Genetics." *The New York Times*. 29 May 1984. P. A1.

Kolata, Gina. "Studying Learning in the Womb." *Science*. Vol. 225: 302.

MacIntrye, Alysdair. "Seven Traits for the Future." The Hastings Center *Report*. February, 1979. p. 5ff.

McCormick, Richard. "Notes on Moral Theology." *Theological Studies*. March, 1979.

McDonald, Kim. "NIH Approves Genetics Experiment Similar to One that Judge Barred." *The Chronicle of Higher Education*. 13 June 1984, p. 7.

McDonald, Kim. "Rapid Growth Genes Could Yield 'Super Livestock'." *The Chronicle of Higher Education*. 8 February 1984. P. 1.

Parker, Philip. "Surrogate Motherhood: The Interaction of Litigation, Legislation, and Psychiatry." *International Journal of Law and Psychiatry*. Vol. 5: 341.

Rahner, Karl. "Christianity and the 'New Man'." *Theological Investigations*. Vol. V. Baltimore: Helicon Press. P. 135ff.

Rahner, Karl. "Experiment: Man." *Theology Digest*. February, 1978.

Rahner, Karl. "The Problem of Genetic Manipulation." *Theological Investigations*. Vol. IX. NY: Seabury Press.

Ramsey, Paul. "Manufacturing Our Offspring: Weighing the Risks." The Hastings Center *Report*. Vol. 8: 7–9.

Ramsey, Paul. "Shall We Reproduce?" The *Journal* of the American Medical Association. Vol. 220: 1346–50 and 1480–85.

Rawley, P. T. "Genetic Screening: Benefit or Menace?" *Science*. Vol. 225: 138–44.

Rifkin, Jeremy. *Algeny*. New York: Penguin Press. 1984.

Robinson, John and Herbert T. Krimmell. "Surrogate Mothers: The Case For and Against." The Hastings Center *Report*. October, 1983.

Roblin, Richard. "Reflections on Issues Posed by Recombinant DNA Molecule Technology." *Annals* of the New York Academy of Science. Vol. 225: 59.

Schneewind, J. B. "Sociobiology, Social Policy and Nirvana." In M. S. Bregory *et al., Sociobiology and Human Nature*. San Francisco: Jossey-Bass. 1979.

Shannon, Thomas A. "The Case Against Test-tube Babies." *National Catholic Reporter*. 11 August 1978.

Smith, David. "Scientific Knowledge and Forbidden Truths." The Hastings Center *Report*. December, 1978. P. 30ff.

Westoff, Charles F. and Ronald R. Rindfull. "Sex Preselection in the United States: Some Implications." *Science*. Vol. 184: 633.